EVOLUTION OF THE UNIVERSE

The Black Hole Theory

Ghassan Hanna Halasa

ISBN: 1523374292
ISBN 13: 9781523374298

To My Loving Wife

Fadia

To My Daughters

Mona

Lina

Zain

Nouf

To My Grandchildren

Francis

Omar

Jamal

Noor

Contents

Preface

When I was young, my father used to show me the night sky, pointing to planets, stars, and galaxies. Since then, the universe has been, for me, a great mystery. I have always wondered how the universe came to be in its present-day balance and about what lies behind what we do not see. I believe that the creation of the universe is a simple process rather than a complicated one that requires the complicated equations of quantum mechanics. I have always believed that using simple classical physics laws is the right way to explain the creation and evolution of the universe. To understand this theory, one only needs basic physics.

The idea of evolution of the universe came to me when I was an undergraduate physics student. Einstein's theory of special relativity and Lorentz transformations were my guidelines in such a way as objects slow down from the speed of light reference frame space expands and mass changes from undefined to defined form which may imply universe creation.

In my early investigation, the following proved facts were the known guidelines to build on:

1. The universe is electrically neutral; both positive and negative charges are equal.
2. Electrons are in continuous motion, orbiting the nucleus.
3. Electrons and protons in atoms are separated without being fused together in spite of the strong electrostatic attraction.

The first point was resolved by assuming that both positive and negative charges were created simultaneously as pairs.

To resolve the second point that electrons are always in motion in atomic orbit; where did the electrons obtain the energy to continue orbiting the nucleus indefinitely? The answer to this question was through the assumption that both protons and electrons bounced away from a reference frame at the speed of light as pairs. The proton traveled to a reference frame that is slower than the speed of light, while the electron traveled to another reference frame faster than the speed of light. Because we observe the proton at rest, it must be in our rest reference frame while the electron stopped in another reference frame, which is in motion in reference to our rest reference frame. This situation can be thought of as sitting on the proton while observing the electron in continuous motion. This assumption explained the electron continuous motion around the nucleus.

The third point was explained by the assumption that the ground state of both the proton and electron is the speed of light. As both elementary particles attract each other in an effort to return to their ground state, the speed of light, their relativistic mass increase was found to be the limiting step that forbids the electron and proton to fuse together. This explained the third point and the secret behind the electrostatic attraction.

By answering the three points above, the mystery of universe creation was resolved.

In further investigation to validate this approach I was faced with two major obstacles in calculating the electron's final speed in orbit around the proton after the formation of the hydrogen atom:

1. If protons and electrons emerged from the speed of light reference frame, does such an inertial reference frame exist at that speed?

2. In their final destinations and from the law of conservation of energy, the proton and electron energies must be calculated. The proton's final destination energy is its mass energy only because its velocity can be assumed to be zero, or at relative rest, implying that its kinetic energy is zero. The electron's final destination energy is its mass and kinetic energies. The major problem in solving this situation is finding the mass of the electron at its final supersonic speed so that its total energy can be calculated right before its attachment in orbit.

The first problem is resolved by using classical physics, where the inertial reference frame can exist at any speed, including the speed of light. The logical answer is that relativistic physics does not rule out the existence of a reference frame moving at the speed of light as long as it exists there all the time. The second problem is resolved by the fact that the electron's mass as measured at rest is, in fact, its mass at the supersonic speed because it is always traveling at that speed when measured.

A direct outcome of this theory is the existence of a parallel universe. This parallel universe cannot be observed from our universe. Black holes are believed to be entrances to the other universe. My theory posits that the other dimensional universe and black holes are the source of all matter.

My theory posits that all elementary and subelementary (such as fermions and fundamental bosons) particles, the sole constituents of the universe, emerged from the speed-of-light reference frame in pairs of positive and negative electrostatic charges. After their exit, they stopped at certain speeds, forming corresponding reference frames housing these elementary particles. Because reference frames are important in understanding this theory, chapter 3 gives it special attention.

The big bang theory, probably the most accepted theory regarding the evolution of the universe, has many unexplained weaknesses. Chapter 4 discusses some of the weaknesses in this theory. The black hole theory is presented as an alternative.

This book is intended for audience with basic classical physics. Although quantum mechanics can be used, but in my opinion, quantum mechanics is used only in situations when classical physics fails to explain certain observed phenomena.

Ghassan H. Halasa

Chapter 1

INTRODUCTION

Throughout history, the origin and the evolution of the universe have been sources of great wonder. How did we get here? How did the universe evolve into the magnificent universe we see as we look at the night sky? We must understand how stars, galaxies, and planets are formed. Was the universe created by a large explosion, as the big bang theory predicts, or was it formed in a continuous, orderly manner? Many questions are associated with the creation and evolution of the major constituents of the universe we see today. The puzzling question that has not been answered is, "Why is the universe, as a whole, electrically neutral?" In other words, why are the positive and negative charges in the universe exactly equal, making this universe electrically neutral? Many answers have been put forward, but none is convincing. Other important questions are, "Why did stars in their early formation possess energy in the form of intense heat? What is the source of this huge heat energy and where did it come from?" The early heat associated with the dense matter in the big bang theory is not a very convincing answer to this puzzling question. Even so, where did the dense matter come from? Other questions include "Why are electrons and protons, the constituents of atoms, different entities, and why electrons are in continuous motion while protons and atoms are at rest?

Why electrons do not spiral down into the nucleus because of the strong electrostatic forces according to Coulomb's force law?" Many more questions arise that have never been fully answered in simple classical terms.

It is observed that stars in their very early formation consist mainly of hydrogen atoms. In this regard, we can conclude that the hydrogen atom is the first element created, and it is definitely the origin of the universe. In a later stage, hydrogen atoms fuse to form helium and other heavier elements. In searching for the universe's evolution, hydrogen atoms must be the primary focus of the search.

Aristotle said that the universe was always there and the earth was the center of the universe. Nicolaus Copernicus formulated a heliocentric model of the universe, which placed the Sun, rather than the Earth, at the center. Sir Isaac Newton combined the contributions of Copernicus, Kepler, Galileo, Descartes, and others, setting the basis of the modern classical mechanics. Newton removed the last doubts about the validity of the heliocentric model of the universe. He formulated the universal gravitational law that explains the precise, delicate structure of the universe.

At the end of the nineteenth century, James Clerk Maxwell gathered the electromagnetic laws of Gauss, Faraday, and Ampere. These electromagnetic laws were put in the form of four partial differential equations (see appendix A). After separating the variables in these equations, Maxwell ended up with an electromagnetic wave that propagated exactly at the speed of light, with the electric and magnetic fields varying with time while being normal to each other and normal to the direction of the wave's propagation. Interested readers may refer to appendix A for derivations and solutions of Maxwell's equations. The wave velocity in free space was found to be the inverse of the square root of the product of the permeability and permittivity of free space. Because these two constants are absolute and universal, the velocity of light must be an absolute constant in free space. Fascinated with this result, Albert Einstein used this result when developing his theories of special and general relativity by assuming that the speed of light is a unique universal constant.

Alexander Friedmann in 1922 solved Einstein's field equations of the theory of general relativity, showing that the universe is expanding. Einstein refused to accept Friedmann's solution, but later he retracted his opposition and accepted Friedmann finding.

Sir James Jeans, in the 1920s, was the first to introduce a steady-state cosmology theory based on a hypothesized continuous creation of matter in the universe. The steady-state theory is a view that the universe is always expanding but maintaining a constant average density, matter being continuously created to form new stars and galaxies at the same rate that old ones become unobservable because of their increasing distance and velocity of recession. A steady-state universe has no beginning or end in time. Jean's theory was never adopted by the general science community.

In 1927, a Roman Catholic priest, George Lemaitre, independently calculated Friedmann's solution, and again he suggested that the universe is expanding.[1-4] The expanding universe was also supported by Hubble in 1929. Hubble found a correlation between the distance of galaxies and the amount of red shift in the galaxy's light.[5] In 1931, Lemaitre proved that the universe is homogenous.[6] He went further with his predictions, extrapolating backward in time when he found that the matter of the universe would reach an infinite density and temperature at a tiny point in space at a finite time in the past, which he dubbed "Primeval Atom."[7] Lemaitre then theorized that the universe came into being billions of years ago when the hot, dense point in space exploded into our present-day universe. It was Fred Hoyle in a radio forecast who gave the name "Big Bang" in 1949.[3] F. Hoyle, H. Bondi, and T. Gold[8] developed the "steady-state model" as a different approach to Lemaitre's primeval atom theory.[6] They concluded that the cosmological principle was valid for space and time. This theory later was found to be inaccurate.[9]

Hoyle updated the steady-state model by the quasi-steady-state model in 1993.[10] Hoyle said that the universe was created from "entirely extraneous spatial dimension."[11] Alan Guth proposed his inflationary theory in 1981. This theory postulates that following the big bang, in the early moments of the universe, an extremely rapid expansion of the universe occurred, driven by a false-vacuum state due to repulsive gravitational action.[12,13] Eric Lerner[14] presented evidence that the big bang theory was contradicted by observations and that another approach, "plasma cosmology," which hypothesized a universe without beginning or end, far better explained what we know of the cosmos. Ned Wright, a strong supporter of the big bang theory, refuted Lerner's plasma cosmology model.[15] Wright is a strong supporter of the big bang theory[16].

In conclusion most scientists do not agree on one strong theory that describes the evolution of the universe. Many agree that the universe started as an expanding point in space.

The big bang theory answered partially the question of the evolution of the universe by stating that it began in a state of very high density and extreme heat 13.7 billion years ago and has been expanding since that particular instant that marked the origin of the universe[6]. "The big bang is the generally accepted cosmological theory; the incorporation of developments in elementary particle theory has led to the inflationary-universe version. There is consensus that the big bang theory is not a theory about the origin of the universe. Rather, it describes the development of the universe over time."[9,11] Since 1931, the big bang theory has been accepted by the wider scientific community as the theory of the origin of the universe. This theory is criticized by the wider religious community and few physicists, although it was accepted by the Vatican as being in agreement with the Bible.

The big bang theory may be criticized by the fact that only one short duration explosion occurred during 13.7 billion years. This theory does not predict successive explosions after the first. Another criticism may be added on what happened to the laws of conservation of energy and momentum before and after the big explosion, where and how the dense matter and intense heat came from. Hoyle said that "it must have come from another dimension."[8] He also predicted that because the universe is electrically neutral, the initial and subsequent processes consisted of neutrons that later disintegrated into protons and electrons. This prediction is important to guarantee the universe electrical neutrality that was never explained by the big bang theory.

The big bang theory may be correct in predicting that the universe was created at the time indicated, but the explosion duration of 10^{-36} seconds, when the universe came into being with its complicated structure and balance that we see today, may be too much to digest. This theory may be based on a combination of both science and theology. Lemaitre's analysis is correct except for the explosion part. If we combine the works of Lemaitre's analysis and Jeans's steady-state model instead of the great explosion, then we might come to a compromise model that the universe was created at the time suggested by Lemaitre but was created slowly and continuously. This model may be more realistic, logical, and easier to comprehend.

In conclusion, no solid theory that describes the origin and evolution of the universe is valid enough to be considered. Almost all existing theories agree that space and mass came together at the same time and formed the vast, fascinating cosmos. Albert Einstein once said, "I have a deep faith that the principles of the universe will be both simple and beautiful.[24]" Einstein's theory of special relativity and Lorentz transformations may be used to describe the universe creation and evolution. This theory may describe the evolution of the universe depending on the position and the velocity of the observer. One weakness of the big bang theory is that it does not satisfy the theory of special relativity.

The dense matter occupying zero space as observed from rest reference frame must have been traveling at the speed of light. Let us assume for the time being that an inertial reference frame exists at the speed of light. Let us assume that a mass, as measured by a traveler in this reference frame, has a mass of one kilogram. According to Lorentz transformation equations, the observer at rest find this mass to be infinite, occupying zero space, which is the same as the dense matter specified by the big bang theory. Let us assume that this mass slowed down from the speed of light until it came to rest. This mass now would measure only one kilogram. For this mass to have any value, ranging from zero to infinity, it will act in the same manner. The question is why Lemaitre did choose the infinite mass, or maybe, his choice was based on the assumption that the mass did not change during its excursion from the speed of light. The second possibility is a strong contradiction to the theory of special relativity. At any rate, the infinite-mass choice has never been explained.

The theory of special relativity tells us that traveling masses cannot reach the speed of light because they require infinite energy to reach that speed because of the increased mass as they approach the speed of light. This theory does not rule out the existence of a reference frame that moves at the speed of light if that frame originally existed at that speed and did not accelerate or decelerate to reach that speed. Also it does not rule out the existence of a reference frame at a speed faster than the speed of light, with the understanding that this frame did not accelerate to reach that speed from rest. The reader should be open-minded to accepting the idea that a reference frame could exist at any speed without thinking that the frames accelerated from rest to that speed but existing originally at that speed. Also Lorentz transformations do not rule out the existence of a reference frame at the speed of light. Inside this reference frame, all laws of physics apply although this reference frame cannot be observed from rest. The existence of such reference frames is essential to understanding this theoretical work. According to classical physics reference frames can exist at any speed; the theory of special relativity accepts this model but restricts reference frames from accelerating to the speed of light from faster or slower speeds.

In this book, simple analysis rather than complicated quantum mechanics is used to show that the origin of the hydrogen atom, the simplest form of matter, is created as proton/electron pair in a reference frame at the speed of light. Hoyle[8] introduced two basic ideas:

1. The universe was created at "extraneous spatial dimension."
2. The big bang started as a continuous flow of neutrons, guaranteeing universe neutrality.

It will be shown that the special dimension, the speed-of-light dimension, does exist. Neutrons are nothing more than electrons and protons pairs produced on a continuous basis at a point in space traveling at the speed of light. The electrons and protons are created in pairs such that the electrical neutrality of the universe is satisfied. Some theories predict the existence of tachyons,[17] particles that travel faster than the speed of light. These theories have not been given much attention. In fact, they are viewed as pure science fiction. In this work, it will be shown that electrons and all negatively charged particles are the tachyons that travel faster than the speed of light.

In this work, it is emphasized that a reference frame does exist at the speed of light. At this reference frame, I theorize and show that electrons, protons, and all other elementary particles, the constituents of all matter, are created continuously and as pairs. Again, the creation of proton/electron pair may simultaneously guarantee the electrical neutrality in the universe. The existence of two independent dimensional universes is an outcome of this work: one universe is our universe, and the other is traveling at the speed of light. The electrons and protons come from this other dimension as predicted by Hoyle.

In context of the proton/electron pair production from the speed-of-light reference frame and the evolution of the hydrogen atom, many other physical phenomena are explained. The following is a list of explanations discussed in this book:

1. Electrons can never spiral down to the nucleus.
2. The temperature of newly formed stars can be predicted.
3. The universe is continuously expanding.
4. The existence of another dimensional universe can be proved.
5. Black holes can be explained and are shown to be gates to the other dimensional universe.
6. The existence of partial mass and charge particles can be predicted.
7. The existence of particles with momentum but without mass can be predicted.

Because the big bang theory is the most accepted theory about the origin of the universe, it is important to discuss its validity and its weaknesses. Chapter 3 is devoted to this topic.

In discussing the theory of the evolution of the universe with colleagues and associates, it was found that their major criticism is centered on a misunderstanding of the concept of inertial reference frames in which one exists at the speed of light. Therefore, chapter 4 is devoted to explaining reference frames in general.

In this book: the terms subsonic, sonic, *and* supersonic speeds *are used to mean speeds below, at, or above the speed of light, respectively. Elementary particles refer to electrons and protons, while subelementary particles refer to particles with or without partial mass and charge as compared to protons and electrons..*

Chapter 2

Preliminary Remarks

The earliest atomic model was introduced by J. J. Thompson in 1904. This atomic structure was also called the "Plum Pudding" model. In this model, Thompson suggested that electrons and protons are distributed in a plum pudding–shaped atom. In 1911 E. Rutherford conducted his scattering experiment, in which he concluded that the atom consisted of a solid nucleus containing the positive charges, which are very small in size that was surrounded by a much larger cloud of negative charges. In 1913 N. Bohr introduced his famous atomic structure, in which positive charges, or protons, are located in the nucleus while electrons revolve around the nucleus in circular orbits. Since then, many other models modifying Bohr's atomic model have been introduced. At any rate all the atomic models speak of a neutral atom with a nucleus housing protons and neutrons bound together with a cloud of electrons around it in fixed orbits of much larger radii. The most important outcome in all atomic models is that protons are considered to be at rest because they constitute the mass of the atom and that atoms are at rest whereas electrons are in continuous motion around the nucleus.

Observations have confirmed the electrical neutrality of the universe. Electrical neutrality means that all positive and negative charges in the universe are exactly equal. Does this equality in the infinite universe mean anything? It is definitely a clear indication that there are two possibilities. If the universe was created in its present-day composition of individual elements and their compounds, this creation includes light and heavy elements. The other possibility is that it was created as proton/electron pair attracting each other to form the hydrogen atom. The first possibility is not very likely because it does not explain why stars in their early formation are made largely from hydrogen. As an example, large portion of the Sun's mass is made from overheated hydrogen gas. This leads to the fact that second possibility is more likely to be correct. In this work, I consider that during the early creation of the universe, overheated hydrogen gas was first to be created. Because the hydrogen atom is composed of one proton and one electron and because of the universe's electrical neutrality, then electrons and protons must have been created in pairs at the same time. This assumption is true for two reasons. One electron and one proton attract each other to form a hydrogen atom; this assumption confirms that the electrostatic neutrality of the universe.

It will be shown and proved that electrons, protons, and all elementary particles were created in pairs in a manner similar to a production line in a factory, where the source of this manufacturing process is not visible to us because it is travelling at the speed of light and we are situated at rest.

The big bang theory does not show why hydrogen gas was the dominant element at the time of universe creation, but predicted that it was created with most present day elements and compounds. On the other hand the proposed black hole theory answers this question, in addition to the following questions:

1. Why electrons are in continuous motion, and why they have they never been found at rest
2. Why positive and negative charges attract each other
3. Why the universe is electrically neutral; that is, why negative and positive charges are equal throughout the universe
4. Why some subelementary particles have partial mass and charge.
5. Where the intense heat in newly formed stars comes from
6. Why the universe is still expanding
7. What black holes are

Matter consists of atoms; atoms consist of electrons, protons, and neutrons. To investigate how matter was created, our attention must be directed toward how electrons and protons were created. The observed fact that the universe is electrically neutral, indicating that positive and negative charges are strictly equal. This means that for every proton, one electron exists and vice versa. This equality leads to the conclusion that positive and negative charges in the universe are strictly created in pairs. All atoms without exceptions are electrically neutral, meaning that electrons and protons are exactly equal in number in all atoms, whatever their size, resulting in an electrically neutral universe. Electrons and protons must have been created in pairs; otherwise, a slight difference in the number of both electrons and protons in the universe would exist.

The Rutherford atomic model defines atomic structure as a stationary, solid nucleus with a cloud of electrons moving in fixed orbit around the nucleus. This brings us to the fact that electrons are always in continuous motion and have never been found at rest whereas protons can be found at rest, such as in an ionized hydrogen atom.

In fact, a researcher cannot physically see an electron. To see an electron, a photon must impact an electron and be reflected to the eye, but the electron absorbs the photon, increasing its kinetic energy and thus increasing its speed. Therefore, it is impossible for the researcher to see an electron or to find an electron at rest. On the other hand, the researcher can easily see a proton at rest. According to Bohr's hydrogen atom model, electrons rotate around the proton (nucleus) in fixed orbits with constant radii and speeds. Is it really the electron that moves around the proton or the other way around? The answer was confirmed by Rutherford in his scattering experiment. According to this model, the electron moves around the proton as cloud. We always speak of the electron energy in orbit, and its quantized energy and electron speed. In fact, we can state without restriction that electrons are always in continuous motion in orbit or in free motion.

Materially, electrons and protons are completely different entities. Their masses, charges, and shapes are completely different. On the other hand, both have masses that can be measured at rest. According to the theory of special relativity, their masses, the space they occupy, and time are functions of their speeds and the speed of the reference frame in which these measurements are taken.

To close this discussion, we can formulate a model that may give a picture of how the universe was created. When we speak of the universe's creation, we speak of mass, space, time, and energy. The following is a brief rundown of the proposed model.

To explain this new approach, let us assume the existence of two inertial reference frames, one at rest and the other traveling exactly at the speed of light. The existence of a sonic reference frame is explained in chapter 3. Let us also assume that both reference frames always existed in that situation at all times. An observer inside the resting reference frame is attempting to observe the inertial reference frame traveling at the speed of light.

Further discussion would lead to the situation in which the resting observer and the observer in the speed-of-light reference frame are unable to detect each other. For the resting observer, the speed-of-light reference frame cannot be detected; in fact, it does not exist.

The speed-of-light reference frame now ejects two particles with momenta of equal magnitudes but opposite directions in its line of motion; one particle will be in the direction of motion of the reference frame while the other particle will be in its opposite direction. According to the theory of special relativity and Lorentz transformations, the observer in the resting reference frame being unable to detect the speed-of-light reference frame now observes two particles emerging from a point of nowhere (a point in space). According to the observer at rest, the two ejected particles move in the direction of the speed-of-light reference frame. From the resting-observer perspective, one particle is traveling at a speed faster than the speed of light while the other particle is traveling at a speed slower than light. In line with the preceding discussion that the electron is always in continuous motion, assume that the particle with a speed greater than light is the electron and that the other particle is the proton. Both the electron and the proton emerged from the speed-of-light reference frame; consequently, the ground state for the electron and the proton is the speed of light. Electrons and protons tend to attract each other to recombine or to return to their ground state; this is basically the nature of electrostatic attraction. According to Lorentz transformations, the masses of both particles become very heavy and eventually infinite near and at the speed of light. Thus, both particles cannot achieve the speed of light because infinite energy is not available to achieve this speed. Therefore, the electron/proton pair production is a one-way process; they can exit the speed-of-light reference frame but cannot return to it. The two particles attract each other in an effort to return to their ground state up to a fixed distance, which can be defined as the atomic radius. This is how the hydrogen atom was created.

Further analysis shows that the mass of the electron is multiplied by a negative imaginary number, i (the square root of -1) because of its supersonic speed. It will be also shown that the negative sign indicates the electronic charge while the imaginary number indicates that materially, the electron is of a different mass form (entity) than the proton, where both electrons and protons are made of two different materials or entities.

It will also be shown that the speed of light exit process is a random one where particles can exit in any direction and it is not restricted to the line of motion of the reference frame. It will also be shown that this random exit process leads, in addition to the creation of protons and electrons, the creation of particles with partial mass and charge or even with tiny mass and momentum if the exit is nearly normal to the direction of motion of the reference frame.

To understand this theory, the reader must be familiar and totally convinced with the following facts:

1. Because of the electrical neutrality of the universe, electrons and protons are created simultaneously in pairs.
2. Electrons are always in continuous motion while protons can be at rest.
3. Electrons and protons materially are of two completely different forms or entities.
4. Inertial reference frames can exist at any speed: below, at, or above the speed of light.

5. For an observer in an assumed resting inertial reference frame (there is no absolute rest reference frame), he or she cannot detect an inertial reference frame traveling at the speed of light. It will be shown later that the speed of light reference frame exists in another dimension.

The following are some highlights of new findings because of this new theoretical approach:

Electrostatic Forces

The electrostatic attraction between electrons and protons is not more than the tendency for the electrons and protons to return to their ground state, which is the speed of light.

Common sense tells us that an electron and a proton should spiral downward and fuse in the hydrogen atom instead of keeping a constant distance from each other. For both elementary particles to fuse together, they must return to their ground state, which is the speed of light. This action requires both particles to have enough energy. According to Lorentz transformations, as the two particles get closer to the speed of light, their masses increase until they exhaust all the energy they possess. The orbital distance in the hydrogen atom creates some sort of balance between available energy and electrostatic attraction. According to Lorentz, mass transformation particles can easily exit the sonic system but cannot return to it. The electron/proton pair production at the speed of light is an accumulative process.

Universe Expansion

Many theories and experimental observations confirm that the universe is expanding. Some theories go further by predicting universe expansion but with a constant density; that is, more stars are created. The big bang theory predicted the universe's expansion only as an aftermath of the original explosion.

In the electron/proton pair production assumption, the proton came to rest while electron stopped at a fixed speed governed by the initial exiting momenta of both particles. The proton's resting speed is only a relative velocity, meaning that the proton may be continuously slowing and that the electron is speeding up. According to Lorentz transformations, space expands as both particles speed away from the speed of light. This is a strong support of universe expansion.

As the electron/proton move away from the speed of light, the expansion of space must be coupled with a reduction in mass. Many observations confirm the universe's expansion, but nothing has been said about mass reduction. Common sense indicates that if the mass of everything in the universe is reduced, then it cannot be detected because the measurement instruments' scales are also reduced.

Calculation of the Electron Speed

My theory assumes that at the speed of light, the electron and the proton exited at equal momenta. The electron ejected in the direction of the speed of light obtaining a speed higher than light, while the proton obtained a speed slower than light. The reader may refer to reference 18 for particles travelling faster than light "tachyons". I calculated the speed of electron in its own reference frame and found it to be 1.8175×10^{10} m/s. According to the Lorentz velocity transformations, the speed was calculated according to rest reference frame and was found to be 9.9×10^6 m/s. This speed is slower than the speed of light. The electron speed as measured from rest is in the order of Bohr's electronic speed in orbit, which is 2.2×10^6 m/s. The subsonic electron speed is a good reason why nobody ever suspected that the electron has a speed that is faster than light.

High Temperatures in Newly Formed Stars

It has been observed that early in their formation, stars consist of hydrogen at very high temperatures. A question may arise as to where and how did this great heat energy comes from. The laws of conservation of energy and thermodynamics state that this energy must have come from somewhere. The big bang theory predicts that this heat was associated with the dense matter right before the original big explosion. The big bang theory does not calculate or predict the amount of heat that came with the early explosion as a fixed value.

This proposed theory of the universe's evolution predicts that the amount of heat and temperature on stars in their early stage of formation. The speed of the electron was calculated to be 9.9×10^6 m/s, which is little faster than the electron speed as calculated in Bohr's hydrogen atomic model, which is 2.2×10^6 m/s. The kinetic energy equivalent to the difference between the two speeds is assumed to be energy that has been converted into heat energy. The temperature associated with the newly formed hydrogen gas is calculated and found to be 1.58×10^6 °C, which is comparable to the temperature found in the Sun's core.

Subelementary Particles

In the electron/proton pair production assumption, both electrons and protons exit the speed of light in opposite directions in the line of motion of that speed. Why the exiting process in the direction of the speed of light direction of motion? Why not in all directions? In fact, the particles' exit could be a random process in which particles may exit in all directions but still do so with momenta of equal magnitudes but opposite directions. The velocity component in the direction of the speed of light may result in pair production of particles with partial mass and partial charge as governed by the cosine of the exit angle they make with the speed-of-light line of motion. If for example the exit is normal to the direction of the speed of light, then there will not be a velocity component in the direction of the speed of light.

Consequently, the exited pair will have no mass but will definitely have momentum.

Existence of Another Dimension

For a reference frame traveling at the speed of light, according to Lorentz transformations, an observer at rest finds the traveling reference frame to occupy zero space with infinite mass and infinite time, which is, by all means of measurements, according to rest observation, nonexistence. On the other hand, an observer in the traveling reference frame would also observe that he or she is at rest while the rest reference frame is traveling away at the speed of light. In a similar analysis, the resting reference frame would appear to the traveling observer on the speed-of-light reference frame to also be nonexistence with zero space, infinite mass and infinite time. The observed infinite mass as observed by the sonic traveler may be the mass of our universe that shrunk to zero volume. Also the infinite mass as observed from rest could also be the mass of another universe that cannot be observed from rest. Therefore, both reference frames must be two universes with comparable masses and sizes existing in what we call different two dimensions; both universes cannot be mutually observed.

The other dimension is the source of the electron/proton pair production in this universe.

Black Holes

For rest observer, the speed of light reference frame has zero distance in the direction of the speed and infinite time. Velocity is defined as distance divided by time, resulting in zero velocity inside the sonic reference frame. Kinetic energy is a function of velocity. The kinetic energy in the sonic reference frame as observed from rest is then zero. It is easy to show that since kinetic energy is zero the rest observer cannot detect any form of movement or energy in the sonic system. This means that the rest observer cannot detect any motion in the traveling reference frame. Because light is a form of energy, the sonic system is observed as dark and dead. The infinite mass observed in the sonic reference frame gives this speed-of-light reference the quality of possessing infinite gravitational force because of the condensed mass. This is the same quality observed in black holes.

Because the speed-of-light reference frame exists in another dimensional universe, as defined in the previous section above, black holes could be gateways or entrances to the other dimension. It will be shown that black holes are simply the factory that produces all elementary particles in this rest universe.

Chapter 3

Reference Frames

Many theories such as the big bang and the steady-state theories have been put forward to explain the creation and evolution of the universe. The science community has failed to accept these two theories, and all others, as strong, reliable theories that fully explain the evolution of the universe over time. In this book, the new theory introduced to explain the evolution of the universe will, hopefully, give a better explanation than its predecessors have. In this theory, the evolution of the universe is based on the assumption that this universe was created from an inertial reference frame traveling exactly at the speed of light. In discussing this theory of evolution of the universe with colleagues, associates, and friends, it appeared to me that almost all objections and criticisms were based on the lack of understanding of the true meaning and behavior of inertial reference frames. Therefore, this chapter is devoted to the explanation of reference frames from both classical and relativistic points of view.

To understand this evolution model, the reader must be well acquainted with the definition of inertial reference frames and Lorentz transformations. In this chapter, reference frames, in general, and inertial, in particular, are discussed. It will also be shown that reference frames can exist at all speeds including slower than, at, or faster than the speed of light. The reader may refer to Appendix B for derivations of Lorentz transformations that are essential to understanding this work.

3.1 Reference Frames

Wikipedia defines a reference frame as "it consists of an abstract coordinate system and the set of physical reference points that uniquely fix (locate and orient) the coordinate system and standardize measurements."[21] A reference frame may be regarded as a closed, rigid environment that independently exists with reference to its surroundings. It may be any size without any restrictions or limitations. An example of a reference frame is the world we live in; on Earth, we can define a set of coordinates and time and can take all kinds of measurements for which all laws of physics are valid. A moving train carriage is another example of a reference frame: a traveler on the train can set his or her own coordinates and time during which experiments can be carried out to validate all laws of physics. Any reference frame could be traveling at some speed in reference to certain fixed position. The reference position could be called rest, or zero-speed reference. If this reference position is traveling, we can specify that the reference position is moving at zero speed. A constraint must be imposed on a reference position such that it is a nonaccelerating system. Other moving reference frames' speeds can be measured in reference to this reference position.

For an accelerating reference frame, a traveler onboard would know if an exerted external force is acting on such a system. For example, if a force is exerted inside this accelerating reference frame, the net force is the sum of internal and external forces. The traveler in such a reference frame can measure precisely the value of external force and his or her system's acceleration.

According to Galilean and Newtonian classical physics, the speed of a reference frame could be at any value without limitations. The speed could range between zero and infinity. This speed could be below at or above the speed of light. According to relativistic kinematics, the speed of light is a region that cannot be reached from faster or slower speeds. The reason for this restriction is that according to the theory of special relativity, as mass approaches the speed of light, its value increases toward infinity at that speed. Infinite mass needs infinite force or energy to reach such a speed, which is impossible to obtain. If an inertial reference frame is at the speed of light, it must have been always there, meaning that it did not accelerate to that speed. Both classical and relativistic physics do not rule out the existence of such a sonic reference frame because the system existed there and did not need infinite energy to be in its sonic location.

Although a sonic reference frame cannot be detected from rest, it does not mean that it does not exist.

Reference frames could be moving or at a standstill, or simply at rest; others could be moving but with variable speeds while others could be at constant speed. For an absolute rest reference frame, it can be argued that such a reference frame may or may not exist because rest is only a relative measure. Being on Earth's surface, someone can argue that this is a resting reference frame, although we know that it is not at rest. The Earth is not at rest because it moves around the Sun; the Sun and the galaxy are also moving; therefore, the Earth's surface is not at rest although we may call it rest. Absolute speed would mean the existence of a preferred or absolute frame of reference, but no such thing exists. No one can calculate precisely the speed of the Earth as an absolute measure.

Specifying that the Earth as a zero-speed reference is allowed when measuring the velocity of other moving objects on its surface. Galileo used the Earth as a zero-speed reference when he measured the velocity of an object when he dropped it from the Tower of Pisa. A police officer measures the speed of a speeding car while sitting in a parked car. The officer can argue that the speeding car has broken the speed limit in reference to his or her parked car. The officer is in one reference frame, which can be called "at rest," while the speeding car, with all its passengers, are in another reference frame at a speed as measured by the police officer. The driver of the speeding car may argue that his or her car was at rest and the officer's car was speeding away in the other direction. In this last argument, the driver considered him- or herself as being at rest. By this argument, the driver considers his or her car as a reference point of measurement. Of course, the judge refuses the driver's argument and finds the driver guilty as charged. The reason for this ruling is because the officer used the Earth as the zero-speed reference because his or her car was at rest in reference to Earth. On the other hand, if both the driver and the officer are living in Jetsons' space age, where the Earth is not there as a reference; driver's argument would now be accepted, and the judge would rule not guilty. In conclusion, for two reference frames traveling at different speeds, either reference frame could be used by the other as the reference point of measurement.

For two reference frames with two different speeds, a traveler on the first taking measurements inside the second system finds the measurements that he took do not match the measurements as taken from inside the second system. To clarify this point, assume two airplanes are traveling at two different constant speeds. The two planes now are at two different reference frames. An observer in airplane one bounces a ball on the floor; he finds the ball tracing a vertical path. The observer on the other airplane finds the ball to trace a triangular path. The same could be said about other physical measurements. The relationship between corresponding measurements taken by any two observers of each other in two different reference frames with two different speeds is called transformation. In classical physics, it is called Galilean or Newtonian transformation; in relativistic kinematics, these are called Lorentz transformations.

3.2 <u>Inertial Reference Frames</u>

An inertial reference frame is defined as a rigid system of coordinates within which the law of inertia applies without being subjected to an external force. In general, an inertial reference frame may also be defined as a nonaccelerating reference frame that is a reference frame traveling at a constant velocity. If a person lives inside a closed box, called an inertial reference frame, which is traveling at constant velocity, he or she would not know if the box is at rest or moving. Inside this reference frame, all laws of physics apply to the traveler regardless of the value of the constant speed of the box. If this box is our universe, then we are in an inertial reference frame not knowing the constant speed of this system. Being on the Earth's surface, we believe that we are at rest when, in fact, we are moving. Almost all physics experiments of motion consider Earth as an inertial reference frame.

A reference frame may be considered as a set of coordinates and time of an independent, rigid environment. Such a system could be a closed train carriage, an airplane, a spaceship, and so on. For a reference frame to be an inertial one implies that no external force is applied to this system. A traveler on such a reference frame would not know if he or she is at rest or moving with constant velocity, in addition, he will not be able to measure his own speed without a reference point. The traveler on such a reference frame can do all kinds of experiments onboard, where all laws of physics apply. Such a reference frame is considered as completely independent system.

From inside an inertial reference frame the speed of such a system can never be measured as an absolute measure. The speed can only be measured in reference to another reference point.

Speeds and velocities are always relative measures. For example, if an observer O_1 is traveling on an airplane with constant speed, v_1, in reference to a stationary observer, O_3, on Earth's surface. A second observer, O_2, on another airplane is traveling with constant speed v_2, again in reference to observer O_3. The three observers are illustrated in figure 3.1.

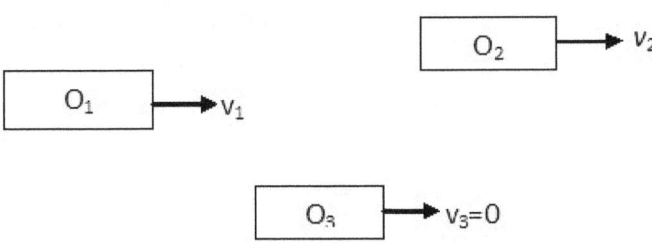

Figure 3.1: The three reference frames

In this example, assume that all three observers measure each other speeds. The three observers are situated in three different inertial reference frames. Each observer can affirm that he or she is at rest in his or her own reference frame. The observer O_3 measures the speeds of O_1 and O_2 and finds them to be v_1 and v_2, respectively. The observer O_2 measures the speeds of O_3 and O_1. Although observer O_3 is at rest, O_2 finds him or her to be moving at a speed of $(-v_2)$ while he or she is assumed to be at rest. O_3 also finds O_1 to have a speed of $(v_1 - v_2)$. Similarly, O_1 measures the speeds of O_3 and O_2 and finds them to be $(-v_1)$ and $(v_2 - v_1)$, respectively. Notice that any inertial reference frame could be a reference to other frames. It is clear that in this example the speed of an object is always a relative measure.

According to Galilean and Newtonian physics, all laws are valid inside any one inertial reference frame traveling at any constant speed. Galilean and Newtonian transformations are applied when observing other reference frames. As an example, a moving body in one inertial reference frame will have different measurements from those measured for the same body from another reference frame. Measurements of motion in different reference frames are always relative measures.

3.3 The Sonic Inertial Reference Frame

In the late nineteenth century, Maxwell used previous electromagnetic laws put forward by Gauss, Faraday, and Ampere to formulate his four famous Maxwell equations (see Appendix A). These equations are four partial differential equations that are functions of space and time. After separating the variables in these equations and with some manipulations, Maxwell found a traveling electromagnetic wave with a phase velocity in free space in terms of two universal constants, equation (A.49) in Appendix A relates them to universal constant speed of light obtained from the solution of Maxwell's equations

$$u = \frac{1}{\sqrt{\varepsilon_0 \mu_0}} = 3 \times 10^8 \ m/s \qquad \text{(A.49)}$$

Where:

u is the phase velocity

$\varepsilon_0 = 10^{-9}/(36\pi)$, Permittivity of free space and

$\mu_0 = 4\pi \times 10^{-7}$, Permeability of free space.

Albert Einstein used Maxwell's result, saying that because the speed of light (the electromagnetic wave) is strictly a function of two universal constants that cannot be changed, then this speed must be also a universal constant. Einstein used Maxwell's findings to put forward his famous theory of special relativity based on the invariance of the speed of light. Lorentz then introduced his famous transformations with Lorentz famous factor setting space, time, and mass as a function of the speed of the reference frame. The derivations of the Lorentz transformations are available in Appendices B and C. Einstein's theory of special relativity was the beginning of new era in physics that changed the understanding of Newtonian and Galilean classical physics. This new approach showed that mass, space, and time, which are known to be invariant in classical physics, are now variables as observed from different reference frames with different relative velocities.

According to classical physics, inertial reference frames can exist at any speed without restrictions. The assumption that Einstein's theory of special relativity and Lorentz transformations changed this classical outlook that inertial reference frames cannot exist at the speed of light is a mistake. **Special relativity deals with objects' behavior only when observed from rest as they approach the speed of light.** Reference frames can be at the speed of light, but the resting observer cannot detect them. Sonic reference frames are considered residents at the speed of light at all times. In general, no law in physics denies the existence of sonic reference frames from classical or relativistic points of view.

In the big bang theory, the science community, in general, has accepted the idea that an inertial reference frame does exist at the speed of light. In fact, the dense matter occupying zero space could have only existed exactly at the speed of light right before the big explosion and only as observed from rest. In other words, the dense matter as depicted by the big bang theory must have been traveling at the speed of light. This dense matter could have a finite mass as measured from inside the sonic reference frame. This finite mass appeared as infinite only as observed from rest. During the big explosion, the dense matter slowed, causing the matter to disperse in a quickly expanding space. When the dense matter slows only slightly from the speed of light, its mass will become finite and its volume will inflate in a quickly expanding space. This is how Lemaitre predicted that through one single explosion the universe was created. In the application of Lorentz transformations, the process of finite mass formation occurred with only a slight reduction in the dense mass speed from the speed of light. Those who believe in the big bang theory must accept the idea that a Galilean reference frame can exist at the speed of light despite such a reference frame being undetected by an observer at rest. This does not change the fact that no reference frame can accelerate from rest or decelerate from a speed faster than the speed of light to that speed. For a reference frame to exist at the speed of light, it must have been created and have existed always there. The reader is reminded that the inhabitants, if any, inside this speed-of-light inertial reference frame enjoy normal life and all physical laws.

3.4 Measurements Near or at the Speed of Light

Assume that a spaceship is traveling at a constant speed, u, in reference to an observation station on Earth set to monitor measurements on the spaceship. This, in fact, represents two inertial reference frames: the spaceship reference frame, which is traveling at constant speed u, and the observation station, which is set to have a speed of zero, the reference measure. For now, let u be little slower than the speed of light. The observation station takes measurements inside the spacecraft and compares them to measurements taken from inside the spaceship or when it was at rest. The observation station finds that:

1. the length measure inside the spaceship along the direction of motion of the spaceship has shrunk to L' from a length L measured when the spaceship was at rest, L' is calculated according to Lorentz transformation equation (B.13) derived in Appendix B:

$$L' = L\sqrt{1 - \frac{u^2}{c^2}} \qquad (3.1)$$

Equation (3.1) shows that the length measure is reduced as the speed (u) increases near the speed of light.

2. the mass in the spaceship has increased to m' from the mass m when the spaceship was at rest. m', as in equation (C.6) derived in Appendix C:

$$m' = \frac{m}{\sqrt{1 - \frac{u^2}{c^2}}} \qquad (3.2)$$

3. The time dilates to T' from T when the spaceship was at rest according to equation (B.18) derived in Appendix B:

$$T' = \frac{T}{\sqrt{1 - \frac{u^2}{c^2}}} \qquad (3.3)$$

where all the primed terms in the preceding equations are as measured by the observation station at rest, the unprimed terms are as measured by the traveler on the spaceship or as measured when the spaceship was at rest, u is the spaceship speed, and c is the speed of light.

Let us now extend the earlier assumption of the spaceship's speed, u, to be equal the speed of light. This assumption does not mean that the spaceship accelerated to the speed of light but that it has always existed at that speed. The observation station finds the following measurements in agreement with equations (3.1), (3.2), and (3.3) when $u = c$:

$$L' = L\sqrt{1 - \frac{u^2}{c^2}} = 0 \quad (3.4)$$

$$m' = \frac{m}{\sqrt{1 - \frac{u^2}{c^2}}} = \infty \quad (3.5)$$

$$T' = \frac{T}{\sqrt{1 - \frac{u^2}{c^2}}} = \infty \quad (3.6)$$

In the development of Lorentz transformation equations, as derived in Appendix B, the x-coordinate in the rest reference frame and the x'-coordinate in the traveling reference frame were assumed to coincide. According to equation (B.9) in Appendix B,

$$x' = \frac{(x - ut)}{\sqrt{1 - \frac{u^2}{c^2}}} \qquad (B.9)$$

a point of discontinuity exists along the x'-coordinate, at u = c, or when the traveling reference frame is exactly at the speed of light. Because of this discontinuity point and at this speed the x'-coordinate in the traveling reference frame is not parallel or coincident with the x-coordinate any longer, but it may assume any direction independent of the rest x-coordinate. As the speed of this reference is increased to more than c, then the coincidence of the two coordinates resume their parallel incidence. The discontinuity exists only at u = c. This is illustrated in figure 3.2. As a consequence, the x'-axis could take any direction such that it may be coincident both the y- or z-axis in the generalized three-dimensional coordinates. This result could be used to show that the volume, rather than the x'-coordinate only, shrinks to zero at the speed of light as observed from rest, which is an indication of space rather than the x'-coordinate contraction.

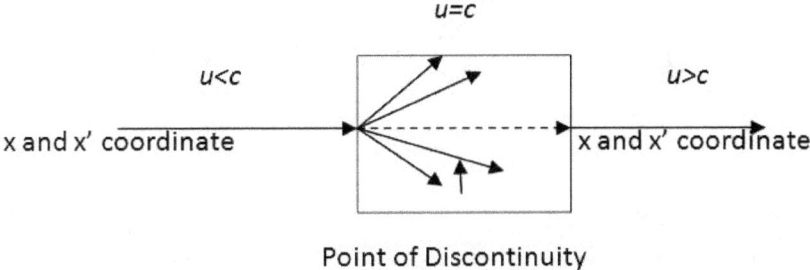

Figure (3.2) The x and x'-coordinate point of discontinuity

The following analysis shows how the velocity measure in the sonic reference frame behaves as measured from rest.

In general, the velocity of a body, v, covering a distance, Δx, in a time, Δt, is defined as

$$v = \frac{\Delta x}{\Delta t} \quad (3.7)$$

At the speed of light, c, all terms in equation (3.7) are primed. Plugging the results in equations (3.4) and (3.6) into the primed (3.7), where $L' = \Delta x'$, $T' = \Delta t'$ and taking the limit,

$$v = \lim_{\Delta t' \to \infty} \left(\frac{\Delta x'}{\Delta t'} \right) = 0 \qquad (3.8)$$

Notice that $\Delta x'$ in equation (3.8) is zero at $u=c$. Equation (3.8) indicates that the resting observer in the observation station finds the velocity of all objects on the spaceship traveling at the speed of light to have zero velocity or, in other terms, motionless. It follows that the resting observer does not detect any kinetic energy inside the sonic inertial reference frame. As a result, the resting observer finds the inside of the sonic reference frame to be dead and motionless. This is a very important result because the observer at rest does not detect kinetic energy inside the sonic reference frame. Photon emission, in general, is associated with molecular motion and collisions that cause electron excitations and electromagnetic emissions. As a result, no light energy is visible inside the sonic reference frame.

In general, the sonic reference frame appears to the rest observer as a dead and dark point in space. Because of the infinite mass in the sonic reference frame, extremely strong gravitational pull is also associated with this sonic reference frame. It will be shown later that this is characteristic of black holes.

3.5 <u>The Other Dimension</u>

From the discussion and analysis in the previous section, several important results could characterize measurements made on the sonic inertial reference frame as measured from rest. The results can be summarized as follows:

1. The volume of the sonic frame is zero.
2. The mass of the zero volume sonic reference frame is infinite,
3. Time dilates to a complete stop.
4. No energy can be observed inside the sonic inertial reference frame.

Because no restrictions were made on the sizes of the inertial reference frame, let us now extend our previous assumption of spaceship and the resting observation post such that the sizes of the two assumed reference frames are very large or approaching infinity. The rest reference frame now is our universe; the sonic traveling inertial reference frame is another rigid system of infinite size to be investigated.

Based on the analysis in the previous section, the rest observer will not be able to detect the sonic reference frame; in other words, he or she would measure zero volume and infinite mass and could not detect any movement or motion inside the traveling reference frame. In fact, the observer in the observation post would see a dark point in space with infinite mass.

Similarly, an observer in the speed-of-light reference frame would find his or her inertial reference frame to be at rest and our universe traveling away at the speed of light but in the opposite direction. An observer in the sonic reference frame would find that his or her reference frame to be a Galilean one enjoying all laws of classical physics while our universe is only a dark point in space with infinite mass. Let us now analyze two important results:

1. As measured from the inertial sonic reference frame, the infinite mass is not more than the huge mass of our universe because our universe appears to the sonic observer as being condensed in zero volume.
2. The shrunken volume tells us that our universe cannot be observed from the sonic inertial reference frame. This indicates that our universe appears to exist in another dimension to an observer in the sonic reference frame.
3. The zero energy appears to the sonic observer as a dark spot, or a "black hole."

Similarly, the resting observer obtains the exact same results when taking measurements of the sonic inertial reference frame. The infinite mass in zero space measured inside the sonic frame as measured from rest could be regarded as another universe existing in another dimension.

3.6 Black Holes

From the preceding discussion about the sonic reference frame, observation of that zone from rest is characterized by

1. zero-volume space;
2. an invisible existence of another dimension, which is an infinite-mass universe reduced at a point; and
3. energy that cannot be detected, thus no light energy can be visible in that zone.

In conclusion, nothing can be observed in the other dimension.

The *Wikipedia* encyclopedia defines a black hole as "[a] **black hole** is a mathematically defined region of space-time exhibiting such a strong gravitational pull that no particle or electromagnetic radiation can escape from it."[20] This definition is a striking similarity to that of a sonic reference frame. The strong gravitational pull may be because of the infinite mass of the other dimensional universe concentrated at a point. Of course, no light can escape from it is because from rest, energy cannot be observed in that zone as explained earlier.

In conclusion, a black hole is definitely a sonic reference frame and could be gates to the other dimensional universe.

It will be shown later that charged plasma fly off from black holes just like dust from a speeding car on a desert road. This is not more than the first step in the evolution of the hydrogen atom which is the first step in the evolution of this universe.

3.7 The Supersonic Reference Frame

As mentioned earlier, according to classical physics that reference frames can exist at any speed less than, greater than, or at the speed of light. This is true only when such frames exist at a corresponding speed, and they do not pass the speed of light barrier. On the other hand, objects can slow or speed up from the speed of light because according to equation (C.6), their mass is reduced from infinity in the case of slowing down; in the case of speeding up, their mass is also reduced but in the negative imaginary form. Later, that supersonic mass is real and can be measured as a finite mass are discussed. With this in mind, let us assume again the existence of two large inertial reference frames: one at rest and the other, a spaceship traveling at a speed faster than light (supersonic speed). An observer in the resting reference frame would have the following measurements according to equations (3.1) and (3.2) and (3.3), with $u > c$, the length measure as observed from the rest reference frame as observed at rest becomes

$$L' = L \sqrt{1 - \frac{u^2}{c^2}} = iL \sqrt{\frac{u^2}{c^2} - 1} \qquad (3.9)$$

where i is the square root of −1.

According to equation (3.9), the length measure is pure imaginary; the resting observer would measure a pure imaginary coordinate in the direction of motion while other coordinates may have real, imaginary or the generalized complex values. In the development of Lorentz transformations, it was assumed that the x'-coordinate in the traveling reference frame coincides with the rest reference frame's x-coordinate. Both coordinates may combine to form a generalized complex x-coordinate. In the general case, the y- and z-coordinates are real and may be complex, forming a complete, more general complex space. Consequently, and because of the generalized space, the resting observer is able to observe and measure activities in the supersonic reference frame.

The mass in the spaceship as observed from rest according to equation (3.2) with the reference frame speed greater than the speed of light c becomes

$$m' = \frac{m}{\sqrt{1 - \dfrac{u^2}{c^2}}} = \frac{-im}{\sqrt{\dfrac{u^2}{c^2} - 1}} \qquad (3.10)$$

And the time as measured from rest, according to equation (3.3), again becomes

$$T' = \frac{T}{\sqrt{1 - \frac{u^2}{c^2}}} = \frac{-iT}{\sqrt{\frac{u^2}{c^2} - 1}} \qquad (3.11)$$

where i in equations (3.10) and (3.11) are square root of (-1). Equations (3.10) and (3.11) show that the measured mass and time in supersonic reference frame is pure imaginary.

As a consequence of equations (3.9) and (3.11), the velocity inside the supersonic reference frame as observed from rest is defined as

$$v = \frac{\Delta(-ix')}{\Delta(-it')} = \frac{(-i)\Delta(x')}{(-i)\Delta(t')} = \frac{\Delta x'}{\Delta t'} \qquad (3.12)$$

The velocity in equation (3.12) as observed in the supersonic frame by the observer at rest is real. Therefore, the observer at rest can observe all objects' movements in the supersonic reference frame, contrary to no movements in the sonic reference frame. The rest observer can also measure kinetic energy of objects in the supersonic reference frame. Because the velocity in the supersonic reference frame is real, the kinetic energy of objects is real and positive in that reference frame.

Equation (3.10) indicates, according to the resting observer, that masses in the supersonic reference frame are negative and imaginary. If we can show that the masses in this reference frame can be measured as real values, then the negative and the imaginary terms associated with this mass must have different interpretations. It will be shown in the next chapter that the negative imaginary masses in the supersonic reference frame are the electrons. The negative sign associated with the supersonic mass must indicate the electrostatic charge of this mass as compared with the real and positive sign associated with the proton mass. Materially, electrons and protons are two different entities; in other words, the material that makes up a proton is different from the material making up an electron. Therefore, it can be said that electrons and protons are two different entities. Therefore, the imaginary term associated with the supersonic mass is no more than an indication of the entity type of the supersonic masses. Consequently, masses in the supersonic reference frame can be considered as real values. As a result, the imaginary term associated with supersonic masses can be dropped. Supersonic masses are then measured as real values as observed from rest, which is the case when electronic masses are measured in the resting reference frame as real, positive values.

If supersonic masses are real and their velocities are real as observed from rest, then their kinetic energies can be observed from rest as real values.

In general, as observed from rest, the supersonic space, contrary to the sonic reference frame, is active with motion. The space in the supersonic space is the same as the generalized rest space; as a result, all constituents and their activities are completely visible and can be recorded from rest.

Chapter 4

Weaknesses in the Big Bang Theory

"In the beginning, when God created the universe, the earth was formless and desolate...everything was engulfed in total darkness...And God said, "Let there be light,"...God made two larger lights...made the stars" (Gen 1,..14). This is how George Lemaitre, a devoted Catholic priest, believed how the universe was created. He wanted to prove the creation in physics terms. He predicted the age of the universe and said that in an extremely short time, our universe came into being in its present shape and structure through a massive explosion "the big bang." He believed that the explosion occurred in hot, dense material occupying zero space.

Today, the big bang theory is the most accepted theory that explains the origin of the universe because it is the only available valid theory. This theory was great in predicting the age of the universe, but it was not very successful in its other explanations. The regression analysis that predicted the age of the universe is definitely correct, but no explanation was given why and how the bang occurred. It never stated where the dense matter came from, if there was other equivalent dense matter, and why there was only one bang in 13.7 billion years.

The big bang theory stated that hot, dense matter exploded in a big explosion that created our present-day universe. The initial dense matter, as predicted by George Lemaitre's big bang theory, had infinite mass and occupied zero space immediately before the bang.

One of the most important questions to be asked is in what reference frame this dense matter was occupying. If this dense matter was and must be in a reference frame what was its speed? Where and who monitored this explosion. The infinite mass of the hot, dense matter is a striking similarity to equation (C.6) for when $u=c$, which defines infinite mass as being at the speed of light, and to equation (B.13), which also defines a zero space at the speed of light. Therefore, the dense matter in the big bang theory must have been traveling at the speed of light, as viewed from rest right before the explosion. For the big bang theory to be correct, there must be two inertial reference frames: one at rest from which to observe the explosion and one at the speed of light where the explosion took place. The dense matter existed initially in a sonic reference frame. In the beginning (before the bang), the rest reference frame did not exist because there was no one in this space (not created yet) to monitor the explosion. Let us assume that a hypothetical monitoring post was there (at rest) to observe the event. The observation from rest is carried out according to the mass transformation equation (C.6) and to the space expansion equation (B.13).

From the sonic reference frame point of view, the mass of the infinite dense matter could have been either finite or infinite as viewed from inside the sonic reference frame. Finite or infinite masses are both viewed as infinite masses from rest. If the dense matter mass was finite, then this mass will be reduced further through its excursion at a lower speed from the sonic speed toward rest. This situation does not explain our universe formation because this lower mass does not reflect the infinite mass of this universe. If the mass was infinite, which is more likely, it may explain the universe inflation as predicted in the big bang theory. If this is the case, then the dense matter must have occupied huge volume in the sonic reference frame as observed from inside that frame. For a huge volume and mass to slow from the sonic reference frame, it must have taken much longer time than the extremely short time as predicted by the big bang theory. This is an important weakness in the big bang theory that has not been criticized by critics or explained by the author of the theory.

4.1 Big Bang Weaknesses

It was pointed out earlier that the big bang theory is accepted by the science and religious communities in general. This does not mean that it is flawless and fully accepted theory. There are many weaknesses that may classify it as a weak theory. The following subsections points out some of the weaknesses.

A. Scattered Debris

During and after the explosion, planets, stars and galaxies dispersed in the newly formed expanding space. No valid reasons were given on why these planets are spherical in shape, have a spin and revolve around each other in perfect balanced equilibrium. Real explosions do not disperse spinning spherical debris scattered in perfect balanced order.

B. Expanding Space and Decreased Mass

As pointed earlier, the hot dense matter in the big bang theory must have been traveling at the speed of light right before the bang. Immediately after the big explosion, the dense matter slowed down from the speed of light, the matter dispersed in an environment of expanding space forming our huge universe that we see today, but nothing was mentioned about what happened to explosion debris as they leave the speed of light to lower speeds.

The big bang theory confirmed the expanding space after the explosion, which is in agreement with Lorentz space transformation equation. This confirms the indication that the dense matter was travelling at the speed of light before the bang. The problem here is why space expansion is in agreement with Lorentz transformation equations and mass reduction is not? The mass of the dense matter must have been reduced as it slowed down from the speed of light. In other words, the mass of the universe is not dense (infinite) any more reflecting the actual mass of the universe. This is a clear contradiction.

C. <u>Recoil</u>

When the infinite-mass dense matter was ejected from the speed-of-light reference frame, it must have been ejected in a direction opposite to the direction of the speed of light so that it slowed down to rest and stayed in the rest reference frame. What happened to Newton's "for every action there is an equal and opposite reaction"? There must have been an equal reaction in the opposite direction, or in the direction of the speed of light. There is no evidence of such occurrence. At any rate, the law of conservation of momentum indicates that there must have been another action occurring at or above the speed of light. The big bang theory does not say anything about a reaction of equal and opposite momentum. This is a clear violation of the law of conservation of momentum.

D. <u>Source of Heat</u>

No valid and convincing explanation was given in the big bang theory to explain the source and amount of heat energy associated with the dense matter. Just saying "hot" is definitely not enough and not scientific. No valid calculations were given to support the heat associated with stars in their early formation.

E. <u>One Bang Only?</u>

13.7 billion years ago the big bang theory predicted that an extremely heavy and hot matter occupying zero space exploded to form our present space and universe; no reason was given for such a tremendous great explosion. This means that right after the explosion, the universe came into being in split seconds with all planets, galaxies distributed and moving in orderly manner. A question was never answered on why the huge planets did not attract each other through their huge gravitational attraction but chose to move around each other in orderly well defined orbits. All this happened in split seconds.

For the sake of argument let us assume that planets, stars and galaxies where created in today's orderly distribution. What if another bang occurred today; would the new explosion create also well defined orderly universe distribution to infiltrate our universe or would such an explosion with all its planets and galaxies collide with the constituents of the old universe?

In conclusion, the age of the universe part is the only correct part of the big bang theory.

4.2 <u>Need For Another Theory</u>

Because of so many weaknesses and unanswered questions in the big theory, a more logical theory is needed to explain the evolution of the universe over time. The theory at hand suggests that the universe instead of being created at once it was created as electrons and protons flew out from black holes or sonic reference frames. Black holes are found to be gates to another dimensional universe travelling at the speed of light. It will be shown that the final destination of the ejected proton is to relative rest, or to zero speed, while the ejected electron reaches a velocity faster than the speed of light. Electrons continue to keep that speed on a continuous basis. This explains the electron's continuous motion in orbit around the nucleus. An observer at rest cannot detect the point of ejection because the speed-of-light reference frame cannot be detected by the resting observer in compliance with the theory of special relativity and Lorentz transformation equations.

After the pair of elementary particles are ejected, and according to the rest observer, the proton/electron pair seem to emerge from nowhere (a point in space). It will be shown that the newly formed mass of the electron is negative and imaginary whereas the proton mass is real and positive. The imaginary and real masses of the electron and the proton, respectively, indicate that the two masses are of two different entities. The negative sign associated with electron mass may indicate the electronic charge.

The two newly formed particles attract each other because of the electrostatic forces. The attractive electrostatic force may be explained by the fact that two created particles maneuver to recombine in an effort to return to their ground state, which is the speed of light. The two particles are unable to recombine because such an action requires the two particles to return to the speed of light. According to Lorentz transformation equations, each particle needs an infinite amount of energy if it is to reenter the speed-of-light barrier. Both particles stop at Bohr's orbit. It will be shown the final speed of the electron is only little higher than the electron speed in Bohr's orbit. The difference in kinetic energy associated with the difference in speeds between the calculated speed of the electron after its exit and its speed at Bohr's orbit is assumed to be converted into heat. This heat energy is associated with the newly formed proton electron pair in orbit, or the newly formed hydrogen atom. This temperature has been calculated and has been found to be in agreement with the Sun's temperature. This heat energy explains the intense heat associated with newly formed stars.

Chapter 5

Evolution of Elementary Particles

The author believes that the universe was first created not as elements of the larger nucleus but as elementary and subelementary particles. An electron and a proton attracted each other, forming the primitive hydrogen atom. At a later stage, hydrogen atoms fused through fusion reaction, forming heavier elements.

Contrary to the big bang theory, what really happened at the time of the big bang 13.7 billion years ago, or may be earlier; rather than the whole universe came into being, only two particles of opposite and equal momenta exited the speed-of-light reference frame along the axis of motion of the sonic reference frame. Their exit was in opposite directions such that one particle was in the direction of motion of the reference frame while the other was in the opposite direction. The equal and opposite momenta ensure that no disturbance (recoil) occurring at the point of exit. Because the sonic reference frame is inertial from the perspective of this reference frame, mass, time, and length measure are finite **as measured from inside this sonic frame**. All laws of physics are valid inside this sonic system including laws of conservation of energy and momentum. The exit of the two particles from the sonic system is implemented according to laws of classical mechanics as measured from the inside of that reference frame.

The laws of energy and momentum are conserved at the moment of exit.

An observer hypothetically situated at an observation post at rest to observe, monitor, and measure the sequence of events of the two particles' as they exit from the sonic reference frame, up to their final destinations. The hypothetical situation at rest is assumed because the very first particles' exit occurs in zero space. Before the particles' exit, the at-rest observer will not be able to detect the two particles when they are in the speed-of-light reference frame. According to the theory of special relativity, an observer at rest detects the sonic reference frame and its contents as a point in space with zero volume and infinite mass. As the two particles exit the speed-of-light reference frame, the observation post now detects and finds that the forward and reverse particles have a speed faster and slower than the speed of light, respectively. This is because both particles left the speed-of-light reference frame in opposite directions: one in the forward (faster) direction and the other in the reverse direction (slower). Both particles are now visible to the observation post at rest because both have real velocities propagating in the same space, as explained earlier.

After the two particles' exit from the sonic reference frame, they exert an effort to attract each other in an attempt to attach to return to their ground state, which is the speed of light. Because the two particles now are in the subsonic and supersonic reference frames, they cannot reenter the speed-of-light barrier. Their forbidden reentry is because their masses increase as they approach the speed of light again. The heavy masses need very large energy to reach the sonic speed, which they lack. Unable to reenter the sonic reference frame, the two particles reach a certain distance from each other and then stop. In this position, they have exhausted most of the energy they obtained from the sonic reference frame. The supersonic particle is now at supersonic speed while the subsonic particle came to relative rest. The moving supersonic particle stays at a fixed distance from the subsonic particle while retaining its supersonic speed. The only possibility for the supersonic particle to keep a fixed distance and the speed it possesses is a circular path around the subsonic particle. If we specify that the supersonic particle is the electron and that the subsonic particle is the proton, then we can declare the birth of the hydrogen atom. The hydrogen atom is formed with the proton at the center of a circular path formed by the moving electron with supersonic speed.

The preceding sequence of events carried out by the two particles after their exodus from the sonic reference frame may lead to three important conclusions: (1) In their effort to recombine to return to their ground state, it is not more than the electrostatic attraction between the proton and electron. (2) The supersonic mass revolving around the subsonic mass is not more than the formation of the hydrogen atom with the electron revolving around the proton. (3) Their attraction stops at a fixed distance that is limited by their increase in mass, which defines the atomic radius.

At the observation post at rest, an observer detects the formation of a hydrogen atom in an inflated space. From the observation post perspective, the newly formed hydrogen atom and inflated space came from nowhere or nothing, because the sonic reference frame is an abstract location as observed from rest. This may answer the controversial question, "Can something come from nothing?" The answer is yes; something can come from an unobserved location.

The two exited particles are now the electron and the proton. The question now is, "Why have we decided that the electron is the supersonic mass and the proton is the subsonic mass?" The answer is simple; because the electron, because of its supersonic speed, is always in continuous motion and never been found to be at rest, it must be the supersonic particle. The proton can be found at rest; therefore, it must be the subsonic particle.

The evolution of space and the simplest element, the hydrogen atom, is the earliest step toward the universe's formation and evolution. The author believes that hydrogen was the earliest constituent of the cosmos. As a result, the evolution of the hydrogen atom and space is the first step toward the evolution of the universe.

The following is a formal presentation of the theory of the evolution of elementary particles.

5.1 <u>Postulates</u>

To proceed with the development and the evolution of elementary particles, the following postulates may be adopted:

1. Inertial reference frames exist at any speed including slower, faster, or at the speed of light.
2. Inside any inertial reference frame, life is normal, and all laws of physics are valid.
3. Masses can exit the speed-of-light reference frame at faster or lower speeds, but they cannot return to that speed (one-way exit).

According to classical physics, inertial reference frames can exist at any speed. Therefore, postulate one is in line with Galilean and Newtonian classical physics. The theory of special relativity did not rule out the possibility that an inertial reference frame can exist at the speed of light. This theory merely stated that such a reference frame cannot accelerate to that speed, and such a reference frame cannot be detected from rest. This theory of special relativity tells us that a sonic reference frame could exist at the speed of light with the constraint that this reference frame originally existed at that speed, and it did not accelerate from slower or faster than the speed of light. On the other hand, no law in physics excludes the existence of a reference frame above the speed of light. Therefore, postulate one could be adopted.

Postulate two is in line with all classical laws of physics as long as the reference frame is nonaccelerating or inertial.

Postulate three complies with the Lorentz mass transformation equation (C.6), derived in appendix C. Equation (C.6) can be rewritten as

$$m = m' \sqrt{1 - \frac{v^2}{c^2}} \qquad (C.6')$$

Where:

m' is the mass in the sonic reference frame at the moment of departure,

m is the mass as measured in the subsonic or supersonic reference frames,

v is the velocity of the particle, and

c is the velocity of light.

Equation (C.6') state that as the velocity of a particle departs from the speed of light (faster or slower), the particle's mass m is reduced to a smaller mass, as measured from rest. The reader can easily verify that a particle cannot reenter the speed of light reference frame because its mass becomes infinitely large, according to equation (C.6').

Consequently, particles can easily exit the sonic speed in both directions, above or below. Because of its nearly infinite mass, as a particle's speed increases toward the speed of light, that particle is forbidden to reenter that speed. As a result, masses can have an easy transition from sonic to subsonic or supersonic speeds while the reverse is forbidden. In general, the sonic reference frame exit is a one-way process. Therefore, postulate three is valid and may be adopted.

5.2 Evolution of Elementary Particles

Based on the preceding postulates, let us assume the existence of three inertial reference frames:

a. Reference frame R_1 is at rest, a "subsonic" reference frame.
b. Reference frame R_2 is traveling at the speed of light, c, a "sonic" reference frame.
c. Reference frame R_3 is traveling with a constant speed v_3 m/s greater than the speed of light, a "supersonic" reference frame.

The assumed three reference frames have their x-axis coordinates coincide and extend along the same generalized x-axis. The velocities of the last two reference frames are in the positive x-axis. The inhabitants, if any, of the three reference frames are independent of each other, living their own lives unaffected by any outside influence whatever the speed of their own reference frame is, as explained in chapter 3.The three reference frames are shown in figure 5.1.

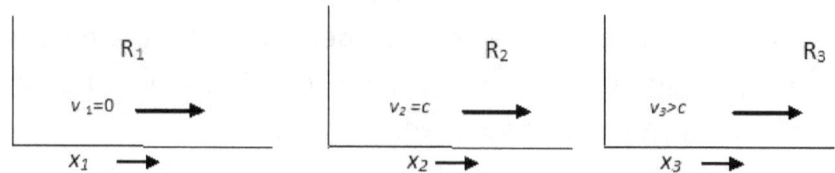

Figure 5.1: The assumed three reference frames; R_1, the resting reference frame (subsonic); R_2, the speed of light reference frame (sonic); and R_3, reference frame at velocity v_3 higher than the speed of light (supersonic)

Assume that a spaceship, or chariot, exists in R_2 and is traveling exactly at the speed of light. The sonic system direction of motion is in the positive x-axis. Also assume that an observation post at rest is situated in R_1 to monitor activities in all reference frames. As explained earlier, the observer in R_1 finds the spaceship with all its constituents as follows:

1. Having infinite mass
2. Occupying zero space
3. The observation post being unable to detect any movements or any signs of life or energy inside the sonic system

No known instruments are available that, when at rest, can detect infinite mass occupying zero volume. In reality, the observer at the resting observation post fails to detect the spaceship in R_2. Although the monitoring post in R_1 cannot detect the assumed spaceship together with all constituents in R_2 the assumed spaceship exist in its own reference frame.

On the other hand, the observation post can detect and take measurements inside the supersonic reference frame as follows:

1. Finite negative imaginary mass
2. Complex space with the imaginary x-axis and the real y- and z-axes
3. Its movements and energy visible can be measured inside it as real values

Because life is normal inside the sonic reference frame as observed from inside that system, a traveler onboard ejects simultaneously two particles p_1 and p_2 along the line of the ship's motion in opposite directions. The ejection is executed according to Newton's laws of motion. p_2 is ejected in the positive x-direction while p_1 is ejected in the negative x-direction. From the perspective of the observation post at rest as being a reference for measurements, particle p_2 is ejected with a velocity, u_2, which is faster than the speed of light, while p_1 is ejected with a velocity, u_1, slower than speed of light. The two particles are ejected with momenta of equal magnitude and opposite directions; the energies of both particles must also be equal such that no residual disturbance occurs at the point of exit. Thus,

$$P_1 = P_2 \quad (5.1)$$

and

$$E_1 = E_2 \quad (5.2)$$

Where:

P_1 and P_2 are the momenta of particles p_1 and p_2, respectively, and

E_1, E_2 are the energies of particles p_1 and p_2, respectively.

The spaceship and the ejected particles example are illustrated in the block diagram of figure 5.2.

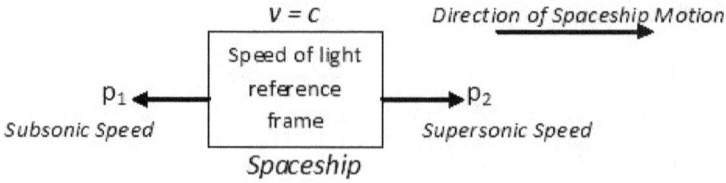

Figure 5.2: Spaceship ejecting two particles

The process of ejection of the two particles is an action that transferred particles p_1 and p_2 from the sonic reference frame to the subsonic and the supersonic reference frames respectively. The ejection of the particles from the sonic reference frame is in line with postulate three, where masses can exit the sonic system.

The observation post in R_1 can now detect and take measurements of the two ejected particle propagating in the newly created common generalized complex space. It is interesting to point out that the observer at the observation post in R_1 is unable to detect the two particles when they were in R_2. The observer at this observation post (R_1) can now affirm that the two particles emerged from unknown point in space and were created from nothing. At this point, we can accept the idea that the universe including both space and matter was created from nothing or unobservable location.

The subsonic and the supersonic particles are propagating in their newly formed, expanding space as observed by the rest observer. The two ejected particles, in an effort to return to their ground state, the sonic reference frame, they attract each other; in such action, the subsonic particle's speed increases while the supersonic particle's speed decreases as both speeds approach the speed of light. The masses of both particles increase as they approach the speed of light. According to postulate three, the two particles can never return to the speed of light. The two particles recombination effort stops at a certain distance. The two particles can never fuse because such an action means that they should renter the speed of light barrier, which is not permitted according to postulate three. The fixed distance between the two particles is established as a balance between energy available with increased mass.

5.3 <u>Evolution of the Hydrogen Atom</u>

The preceding discussion of the ejection of the two equal momentum masses poses a striking resemblance to the creation of the electron/proton pair. The supersonic and the subsonic ejected particles are the electron and the proton, respectively, because of the following:

1. As observed from rest, the ejected supersonic particle, according to equation (3.10), is negative imaginary term. The negative term may be related to the electrostatic charge of this mass. The supersonic particle resembles the electron because of its continuous motion.

2. The subsonic ejected particle's mass is positive and real and attract the supersonic mass; then it must be the proton.

3. The imaginary term associated with supersonic mass could be related to the difference between the texture and the entity of the two ejected masses.

4. The attraction between the supersonic negative mass and the subsonic positive mass, in their effort to return to their ground state, is evidence of the electrostatic attractive forces of the two ejected masses.

5. The two evolved masses cannot fuse because such an action requires that the created masses must reenter the sonic reference frame, which is forbidden according to postulate three.

It can be generalized that the two ejected mass are not more than an electron and a proton. Their attraction up to a certain distance and the electron's continuous motion around the proton are strong evidence of hydrogen atom formation. The fixed radius between the electron and the proton explains why electron does not fuse with the proton in the hydrogen atom disregarding the strong electrostatic attractive forces.

Let us now modify our above assumption of the spaceship. Assume that because of an unknown action at the sonic reference frame, at the other dimension or may be at a black hole, a sonic mass split into two masses. The two masses obtain energy to split and exit their sonic home with equal and opposite momentum along the direction of motion of the sonic reference frame. The two split particles are the electron and the proton forming the earliest hydrogen atom.

In the next chapter that the particles' exit from the speed of light is not always in the direction of motion of the sonic reference frame will be shown. The exit process is rather random and could be in any direction.

Chapter 6

Supporting Analysis

Earlier, it was shown that the hydrogen atom evolved when some mass in the sonic reference frame gained enough energy to split into an electron, which traveled ahead of the sonic system, and a proton, which trailed that system. The exit was implemented on the basis of equal energy and momentum. This equality could be confirmed because no visible energy has accumulated over billions of years of elementary particles emission anywhere. For the two particles to recombine in an effort to return to their ground state, which is the speed of light; they must renter their sonic home. In their effort to return to the sonic speed, both masses become very heavy as they approach the sonic speed barrier, their reentry is forbidden according to postulate three. Consequently, some sort of balance is formed between the two ejected particles' mass increase and the electrostatic attractive forces such that a stable hydrogen atom is formed. As a result, the hydrogen atoms creation continued over billions of years to be cumulative because of the one-way exit, leading to the formation and evolution of this universe. The particles' exit is associated with space expansion.

To prove the validity of this proposed model, mathematical calculations based on this theory must agree with previously obtained results and measurements. It will be shown that the speed of the electron in orbit in the hydrogen atom is only little higher than the speed as calculated by Bohr. The difference between the two speeds is assumed and found to be excess kinetic energy that is converted into heat energy associated with the newly formed hydrogen atom. The temperature rise agrees with temperature found on newly formed stars.

6.1 Energy Conservation in Reference Frames

My theory posits that when the two particles exit the sonic location, their mass, their speed of exit, their energy, and their momenta are equal. After the exit, their masses changed because of relativistic consideration associated with changes in speed and mass. Although their energy and momentum are conserved, energy, although still conserved, I believe that it tends to change into other forms. The energies include kinetic, mass, electrostatic and gravitation.

From the observed fact that the point of exit is unknown, as a result, this point is considered as undisturbed, and therefore, the exit is implemented on momentum and energy equally. This means that at the moment of exit, both particles have equal energy and momentum.

At the moment of exit, the momentum of the forward and reversed particles is given as follows:

$$P_1 = m_1 v_1 = P_2 = m_2 v_2 \qquad (6.1)$$

where P_1, P_2, v_1, and v_2 are the two particles' momenta, masses, and velocities at the moment of exit as measured at the sonic reference frame.

The kinetic energy associated with both particles from the sonic reference frame perspective is given by

$$k_1 = \frac{1}{2}\frac{P_1{}^2}{m_1} = \frac{1}{2}m_1 v_1{}^2 \qquad (6.2)$$

And

$$k_2 = \frac{1}{2}\frac{P_2{}^2}{m_2} = \frac{1}{2}m_2 v_2{}^2 \qquad (6.3)$$

where the subscript 1 denotes the subsonic particle and the subscript 2, the supersonic one. The kinetic energy in equations (6.2) and (6.3) of both particles are equal at the moment of exit, again because no energy accumulation is observed at the point of exit. For equations (6.1), (6.2), and (6.3) to be valid, the particles' exit masses, m_1 and m_2, and velocities, v_1 and v_2, must be equal at the point of exit.

After the particles' exit, the kinetic energies in equations (6.2) and (6.3) are reduced and have changed into other energy forms, such as mass energy and electrostatic attractive energy. At this point, it must be emphasized that their total energy is still conserved.

The subsonic particle or the proton possesses the following energy forms as measured in the subsonic reference frame.

$$E_{mp} = m_p c^2 \qquad (6.4)$$

$$E_{kp} = 1/2\, m_p v_p^{\,2} \qquad (6.5)$$

$$PE_p = \frac{1}{4\pi\varepsilon_0}\frac{q_1 q_2}{r} \qquad (6.6)$$

Where:

E_{mp} is the proton mass energy,

E_{kp} is the proton kinetic energy,

m_p is the proton mass as measured in the subsonic reference frame,

PE_p is the proton electrostatic potential energy,

v_p is the proton speed in the subsonic reference frame,

c is the speed of light,

ε_0 is the permittivity of free space,

q_1 and q_2 are the charges of the proton and the electron, and

r is the distance between the electron and the proton.

The total energy of the proton in its final destination, E_{tp}, is equal to the sum of the three interchangeable energy forms in equations (6.4), (6.5), and (6.6); thus,

$$E_{tp} = E_{mp} + E_{kp} + EP_p \qquad (6.7)$$

Substituting equations (6.4), (6.5), and (6.6) into equation (6.7),

$$E_{tp} = m_p c^2 + 1/2\, m_p v_p^{\,2} + \frac{1}{4\pi\varepsilon_0}\frac{q_1 q_2}{r} \qquad (6.8)$$

Equation (6.8) is the total proton energy in the resting reference frame.

Similarly, the electron has the following energies in the supersonic reference frame:

$$E_{me} = m_e c^2 \qquad (6.9)$$

$$E_{ke} = 1/2\, m_e v_e^{\,2} \qquad (6.10)$$

$$PE_e = \frac{1}{4\pi\varepsilon_0}\frac{q_1 q_2}{r} \qquad (6.11)$$

Where:

m_e is the electron mass as measured in the supersonic reference frame,

E_{me} is the electron mass energy,

E_{ke} is the electron kinetic energy,

PE_e is the electron electrostatic potential energy,

v_e is the electron speed as measured in the supersonic reference frame,

c is the speed of light,

ε_0 is the permittivity of free space,

q_1 and q_2 are the charges of the proton and the electron, and

r is the distance between the electron and the proton.

Similarly, the total energy of the electron in the supersonic reference frame, E_{te}, is also the sum of the mass, kinetic, and electrostatic energies:

$$E_{te} = E_{me} + E_{ke} + PE_e \qquad (6.12)$$

Substituting equations (6.9), (6.10), and (6.11) into equation (6.12),

$$E_{te} = m_e c^2 + 1/2\, m_e v_e^{\,2} + \frac{1}{4\pi\varepsilon_0} \frac{q_1 q_2}{r} \qquad (6.13)$$

The energy in equation (6.2) is the total initial energy, k_1, of the proton at exit, which is totally kinetic energy. As explained earlier, the initial energy of the proton is equal to the initial electron energy, k_2, in equation (6.3):

$$k_1 = k_2 \qquad (6.14)$$

The law of conservation of energy tells us that the particles' energy at the moment of exit must equal their energy at their final destination; this is true because no energy loss is observed during their excursion to their final destinations. Therefore, E_{tp}, the proton's total energy in equation (6.8), must equal the proton's initial energy at the point of exit, k_1, and the electron total energy, E_{te}, in its final destination must also equal the electron's initial energy at the point of exit k_2; consequently,

$$k_1 = E_{tp} \qquad (6.15)$$

$$k_2 = E_{te} \qquad (6.16)$$

therefore,

$$E_{tp} = E_{te} \qquad (6.17)$$

Consequently, the right-hand sides of equations (6.7) and (6.13) are equal; thus,

$$m_p c^2 + 1/2\, m_p v_p^{\,2} + \frac{1}{4\pi\varepsilon_0} \frac{q_1 q_2}{r}$$

$$= m_e c^2 + 1/2\, m_e v_e^{\,2} + \frac{1}{4\pi\varepsilon_0} \frac{q_1 q_2}{r} \qquad (6.18)$$

Where:

m_e is the electron mass as measured in the supersonic reference frame,

m_p is the proton mass as measured in the subsonic reference frame,

v_e is the electron speed as measured in the supersonic reference frame,

v_p is the proton speed as measured in the subsonic reference frame,

c is the speed of light,

ε_0 is the permittivity of free space

q_1 and q_2 are the charges of the proton and the electron, and

r is the distance between the electron and the proton.

6.2 Calculating Electron Speed

After the electron/proton pair exits the speed-of-light reference frame, the electron reaches a final speed in the supersonic reference frame, v_e, while the proton relatively comes to complete stop, or to rest, in the rest reference frame; the proton speed, v_p, in equation (6.18) is then equal to zero. The electrostatic potential energy terms cancel out in equation (6.18) because the electrostatic potential between positive and negative charges is the same as that between negative and positive charges. Equation (6.18) reduces to

$$m_p c^2 = m_e c^2 + 1/2\, m_e v_e^2 \qquad (6.19)$$

Equation (6.19) indicates that the proton's mass energy as measured in the resting reference frame equals the mass energy and the kinetic energy of the electron as measured in the supersonic reference frame. In equation (6.19), that the speeds of the electron and proton are relative is assumed, with the proton's speed as zero.

The final destinations of the electron and the proton are in the hydrogen atom, where the electron moves with supersonic speed around a stationary proton. Two reference frames exist inside the hydrogen atom; the proton resides in the resting reference frame while the electron resides in the supersonic reference frame.

Because the proton has escaped the sonic system and resides in the resting reference frame, its energy, velocity, and momentum are real. As a result, the left-hand side of equation (6.19) is real. The electron's exit is toward the supersonic reference frame; as shown earlier, the electron's supersonic speed is real while the electron supersonic mass is associated with a negative imaginary term. Also as explained earlier, the negative term represents the electrostatic polarity of the electron while the imaginary term represents the characteristic property of the electron mass. Because the electron's mass is measured as a real value in the lab at rest, both the negative and the imaginary terms associated with the electron's mass can be dropped, and the mass can be measured precisely from rest as real value but in the supersonic reference frame. Consequently, the right-hand side of equation (6.19) is also real. Therefore, both sides of equation (6.19) are real values.

It was also shown that the newly formed expanding space for the electron is a semicomplex space while the expanding space for the proton is real; therefore, both spaces define the same mathematically generalized complex space, where the newly formed electron and proton are floating. It was also shown that the supersonic particle, the electron, and the subsonic particle, the proton, can now be observed from rest as real values attracting each other through their efforts to return to their ground state, which is the speed of light.

Relativistic mass transformation tells us that as a particle slows down from sonic reference frame its mass decreases, only as observed from rest. On the other hand, as a particle speeds up from the sonic reference frame the magnitude of this mass also decreases. As for the electron, when it speeded up from the sonic system its mass is reduced and can be measured from rest. The measured mass at rest is its actual mass as measured inside the supersonic reference frame because of its continuous motion. Therefore, the electron mass of 9.10908×10^{-31} kg is the supersonic mass, not the resting mass normally referred to. The actual rest mass of the electron can be found by using the transformed mass m_0 at rest after dropping the negative and imaginary terms:

$$m_0 = m' \sqrt{\frac{v_e^2}{c^2} - 1}$$

$$\approx 9.10908 \times 10^{-31} \times \sqrt{\frac{(1.8175 \times 10^{10})^2}{(3 \times 10^8)^2}}$$

$$= 5.5186 \times 10^{-29} \, kg \qquad\qquad (6.20)$$

Where v_e is the supersonic speed calculated on the next page, equation (6.21)

The true resting mass, m_0, of the electron in equation (6.20) is about two hundred times larger than the electron mass as usually referred to when measured from rest.

The electron must have final speed $v_e > c$ in the supersonic reference frame. From rest, the electron's mass can also be observed and measured as a real, positive value. The measured electron mass of 9.10908×10^{-31} kg at rest is, in fact, its mass in the supersonic reference frame because the electron always exits at supersonic speed. Consequently, the electron mass in equation (6.19) represents its mass in the supersonic reference frame. As a result, all values in equation (6.19) are real and known values except for the supersonic electron speed, which is unknown.

The following are the values of terms in equation (6.19):

m_p is the rest mass of the proton = 1.62752×10^{-27} kg;
m_e is the relativistic mass of the electron
= 9.10908×10^{-31} kg;
c is the speed of light = 3×10^8 m/s, or 1 light-year/year;

and

v_e is the supersonic speed of the electron, which needs to be evaluated.

Rearranging terms in equation (6.19) and setting the speed of light as one light-year per year, v_e, the electron speed in the supersonic reference frame, is found to be

$$v_e = \sqrt{\frac{2(m_p - m_e)}{m_e}} = 60.584 \frac{lyr}{yr} = 1.8175 \times 10^{10} \ m/s \qquad (6.21)$$

The velocity, v_e, in equation (6.21) is the actual velocity of the electron in the supersonic reference frame. As expected, the electron speed is faster than the speed of light.

A question may arise as to why we always measure the speed of the electron as being blow the speed of light and not the speed as depicted in equation (6.21). The answer is simple: the speed in equation (6.21) is measured from inside the supersonic reference frame. The measured electron speed in the resting reference frame requires that the speed in equation (6.21) be transformed to the resting reference frame. Using the Lorentz transformed velocity in equation (B.21) in Appendix B,

$$V = \frac{(V'+u)}{(1+\frac{uV'}{c^2})} \qquad \text{(B.21)}$$

Where:

V' is the velocity of the electron in the supersonic reference frame,

u is the supersonic reference frame speed,

V is the transformed velocity as measured by the observer at rest,

c is the speed of light.

The velocity of the electron, V', is taken to be the same as the velocity of the reference frame u, as calculated in equation (6.21); thus,

$$u = V' = 1.8175 \times 10^{10} \, m/s \quad (6.22)$$

Plugging the calculated speed in equation (6.21) into equation (B.21) and using the figures in equation (6.22), the speed v_e' of the electron as measured at rest will be

$$v_e' = 0.00330 \, l\text{-}y/yr = \mathbf{9.9x10^6 \; m/s} \quad (6.23)$$

The electron speed in equation (6.23) is the electron's observed speed as measured by resting observer. This result is interesting because it is slower than the speed of light as always being observed. The speed of the electron in equation (6.23) is in the order of the electron's speed as calculated in Bohr's hydrogen atom model, which is

$$v_H = 2.2 \times 10^6 \; m/s \quad (6.24)$$

The results from equations (6.23) and (6.24) are clear indicators of the electron and the proton having enough energy to recombine as well as have some excess kinetic energy. In the following, it will be shown that this excess energy is converted into heat.

6.3 Heat Associated with Newly Created Hydrogen Atom

The calculated electron speed in equation (6.23) above is an important result because it is only little higher than Bohr's speed in equation (6.24). It is clear that the newly formed hydrogen atom has more energy than a normal hydrogen atom as presented by Bohr. The excess energy could be calculated as follows:

ΔE_k =1/2 $m_e v e'^2$-1/2 $m_e v_H^2$

    = ½ m_e [$(9.9 \times 10^6)^2$- $(2.2 \times 10^6)^2$]

    =4.244x10^{-17} Joules/one hydrogen atom   (6.25)

Equation (6.25) is the electron's excess kinetic energy associated with newly formed hydrogen atom. The question is what happens to this excess energy and where did it go. How does this excess energy affect the hydrogen atom evolution? The most logical answer is that this energy is converted into heat energy as observed on stars.

The kinetic energy in equation (6.25) is per one hydrogen atom. Two hydrogen atoms attach, forming a hydrogen molecule. Energy per one hydrogen molecule is simply the figure in equation (6.25) multiplied by two

ΔE_{1m}=2 x 4.245x10^{-17}

    = 8.49x10^{-17} Joules/one hydrogen molecule   (6.26)

The energy in equation (6.26) is multiplied by Avogadro's number

$$N_A = 6.022 \times 10^{23} \text{ mol}^{-1}$$

To find the energy in one mole of hydrogen gas, equation (6.26) is multiplied by the above N_A,

$$\Delta E_{mole} = 8.49 \times 10^{-17} \times 6.022 \times 10^{23} = 5.1536 \times 10^7 \text{ Joules/mole} \qquad (6.27)$$

The energy in equation (6.27) is the excess kinetic energy available in one mole of the newly formed hydrogen gas. This excess energy appears as heat energy in the newly formed hydrogen gas.

To convert the energy in equation (6.27), let us assume that the initial hydrogen gas temperature is absolute zero degrees Kelvin. The energy in (6.27) must raise the temperature of the hydrogen gas from absolute zero to some temperature. To calculate the rise in the hydrogen gas temperature above the absolute zero, the following procedure may be followed.

The hydrogen gas specific heat is C_p= 32.58 J/mol-K (at high temperature).[23] The increase in the temperature of the gas from absolute zero temperature would then be

$$\Delta E_{mole} = C_p \times (\Delta T) \qquad (6.28)$$

where ΔE_{mole} is the excess energy in equation (6.27) and ΔT is the temperature rise, in kelvins, associated with the newly created hydrogen gas. Solving for ΔT in equation (6.28),

$$\Delta T = \frac{5.1536 \times 10^7}{32.58} = 1.58 \times 10^6 \, K \approx 1.58 \times 10^6 \, ^\circ C \qquad (6.29)$$

The temperature rise in equation (6.29) is the temperature of the newly formed hydrogen gas. Scientist believed that this temperature is comparable with the temperature found on stars in their early formation.

The temperature rise in (6.29) is enough heat to cause fusion reaction, resulting in the formation of heavier elements in nature[22] associated with further rise in temperatures in stars.

The heat associated with newly formed hydrogen gas is more logical explanation than that explained in the big bang theory.

6.4 Evolution of Subelementary Particles

In the preceding analysis, it was assumed that the x-axis is the direction of the speed of light and is the direction of the velocities of the ejected electrons and protons. A critic may wonder why the x-axis only and not all directions? As a matter of fact, this exit process may be generalized so that the particles could exit in all directions. The ejected particles could exit in all directions lateral to the speed of light direction. The only restriction is that the ejection is in pairs and that the exited particles must have equal and opposite momentum such that no disturbance can be inflicted on the point of exit.

Assume that two particles with equal momentum exit the sonic reference frame at an angle α with the direction of the speed of light, as shown in figure 6.1.

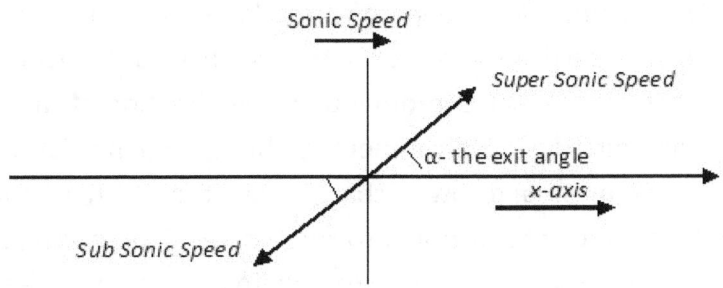

Figure 6.1: Random exiting process lateral to the direction of the speed of light

Figure 6.1 shows a pair of particles exiting the sonic reference frame at angle α with the x-axis, the direction of the speed of light. One particle obtains a velocity component in the direction of the supersonic speed while the other particle has velocity component in the direction of the subsonic speed. The components of both speeds along the x-axis are given by

$$v_x = v \cos \alpha \qquad\qquad (6.30)$$

where v_x is the component of the exiting particle's velocity along the x-axis and v is the actual random exiting speed. The exiting particles obtain speeds slower and faster than the speed of light but with a factor less than one, which is the cosine of the exiting angle. The x-component speed of these lateral exiting particles is less than the particles that form the electron and the proton. Therefore, these particles' momenta in the direction of the speed of light are also less. Consequently, the exiting particles must obtain lower mass and lower charge than the electron and the proton by a factor proportional to the cosine of the exiting angle. These particles may be related to subelementary particles with partial mass and charge. Probably some these subelementary particles are useful in the later formation of multiproton elements.

At a later stage, some of the formed subelementary particles with enough energy may attract each other in an effort to return to the ground state, which is the speed of light. In this action, their masses and charges increase as they approach the speed of light. Through this action, they behave like electrons and protons when they stop at a fixed distance, similar to combination of electron/proton pair forming more hydrogen atoms, thus increasing the population of hydrogen but with less heat energy.

Some of these subelementary particles may depart the sonic reference frame in a direction close to normal to the direction of the speed of light, or at nearly $\alpha = 90°$. Consequently, they have almost no velocity components in the x-axis, the speed of light direction. Then these subelementary particles have no mass but have momentum only. These particles could be the neutrinos that have almost no mass but with momentum.

6.5 Universe Expansion

In the preceding development of the evolution of the hydrogen atom it was assumed that several inertial reference frames in motion with different speeds exist in reference to each other. One reference frame is assumed to be at rest so that the speeds of all other frames can be established in reference to this rest point. The planet we live on was chosen to be the resting reference frame. From Earth, we looked at a hypothetical point that we called the speed of light or sonic reference frame and another reference point we called the supersonic reference frame. The proton/electron pair was ejected from the sonic reference frame. The proton speed decreased until it came to a stop in the rest reference frame; while the electron stopped in the supersonic reference frame.

As the proton/electron pair exit the sonic reference frame, their masses are reduced and the space they propagate in expands as these particles speed away from the point of exit. In this situation a question may arise: "Why did the proton come to a complete stop in the reference frame that we called rest and the electron speed stopped at the specified supersonic reference frame?" The answer most probably is that both particles' motions continued indefinitely. In this case, the masses of both particles are continuously decreasing and the space they propagate in is continuously expanding. This means that the specified rest reference frame is not at rest any more. The increase in space may be regarded as the "expanding universe." The decrease in mass may not be detected in the resting reference frame.

Let us assume that all the protons and the electrons, as we know them have variable masses during their trek from the sonic reference frame and that these masses are reduced to half values as we know them. Is it possible for us to detect this mass loss?

Certainly we would not be able to detect this loss because we and all our measuring instruments and scales are made from protons and electrons that have also had their scales reduced to half. Consequently, we would not be able to detect the mass loss. On the other hand, universe expansion can be detected as found in theories of universe expansion.

Explaining why the proton/electron distance in an atom did not change during this mass loss and expanding space is left. As explained earlier, the atomic distance between proton and electron was fixed because of the particles' energy content that is limited by their increased masses as they approach the speed of light. As explained earlier, the atomic radius is limited by a balance between the electrostatic attraction and relativistic mass increase. Therefore, the atomic size does not change with universe expansion.

It can be stated that the set rest reference frame is not at an absolute rest. As a result, space expansion can be detected; while mass reduction cannot be detected. This may support universe expansion.

Probably the above universe expansion argument may be not very convincing and can be contested. Therefore it is suggested that further investigation is needed.

6.6 Constant Density Universe

In the above development of the evolution of the hydrogen atom, it was assumed that the proton/electron pair production is a continuous process. This elementary particles pair production may be regarded as a continuous and nonstop process resulting in continuous mass creation. It was shown, and in line with postulate three, the production is one way; it is a cumulative process, meaning that those protons, electrons, and all elementary particles are being produced continuously. Hydrogen atoms are being created in a nonstop process; on the other hand, the size of space is increasing; the result may support the constant density universe.

6.7 <u>Particles' Emission Profile</u>

It is assumed that the particles' emission at the sonic reference frame is not implemented in an orderly fashion but, rather, is a random process in which particles fly off in all directions. This process may resemble a speeding car on a desert road. The only constraint is that the emission is carried out in two equal masses with equal and opposite momentum.

The particles emitted in the direction of the sonic reference frame pair up to form electrons and protons. Electrons and protons attract each other, forming overheated hydrogen atoms. For these emitted particles, which make small angles with the direction of the speed of light, they continue to form hydrogen atoms but with less heat energy. Those emitted particles that make angles that are larger but lower than 90 degrees form subelementary particles that have partial masses and partial charges compared to electrons or protons. A lateral emission (at 90 degrees or so) in the speed-of-light direction, the resulting particles do not have mass or charge but momentum.

The location of this emission is the speed of light. It was shown earlier that the other dimensional universe is the speed of light. We can speculate that the source of the particles emission is the other dimensional universe. It was shown earlier that black holes are the entrances to the other dimensional universe; therefore, we can speculate that the source of this sonic dust is black holes.

Chapter 7

Evolution of the Universe

At the beginning of our universe, 13.7 billion or more years ago, as predicted by the big bang theory, it was empty with zero volume, as observed from a subsonic reference frame, the reference frame we live in. The zero volume also stems from the assumption that there was nothing in this universe that can be referred to as a resting reference frame. Through the action of an unknown mechanism in the sonic reference frame, two particles were ejected in the direction of motion of the sonic system. The two particles were ejected with equal and opposite momentum. The equal and opposite momentum did not cause any disturbance at the point of exit. From the exit symmetry, one particle obtained supersonic speed while the other particle obtained subsonic speed. The supersonic particle is the negatively charged electron while the subsonic particle is the positively charged proton.

The opposite polarity charges are because both ejected particles attract each other in an effort to recombine so that they could return to their ground state, which is the speed of light; this attraction is referred to as the electrostatic attraction. To reenter the speed of light and according to Lorentz transformation, both particles need infinite energy because their masses approach infinity as they get closer to the speed of light. The infinite mass limits their recombination; through this process, they stop at a certain distance from each other; this distance is the hydrogen atom radius. The result is the creation of the first hydrogen atom. Simultaneous with the exit process, space expanded as the particles speeded away from the sonic system. The expanded space marked the first space to house the newly formed hydrogen atom; in addition, time slowed to finite values. As measured from rest, the speed of the electron in orbit around the proton is found to be only little lower than the electron speed at its final destination, which is the hydrogen atom orbit. The excess kinetic energy available to the electron in orbit appears in the form of heat that raises the temperature of the newly formed hydrogen gas. More and more overheated hydrogen atoms are created in the newly formed space.

The formed overheated hydrogen atoms attracted each other, possibly by the action of gravitational pull forming small colonies of hydrogen atoms that changed later into overheated stars. The process continues where more and more stars and galaxies are being created in expanding universe.

As a consequence of the universe evolving as laid out earlier, its size increased in a continuous and slow process. Based on the evolution of protons and electrons and the eventual formation of the overheated hydrogen atoms, the origin of all elements in the universe is more logical than is the explosion model presented by the big bang theory.

A question that has baffled scientists, philosophers, and theologians is, "Where did this universe come from?" They all agree on the logical fact that nothing should come from nothing. The answer to this question is simple: the universe came from a point that is not observable to us from a resting observation location.

7.1 Theory of Evolution of the Universe

In the beginning, our universe was a point traveling at the speed of light. Through an unknown action, two particles were ejected in the direction of the speed of light with equal and opposite momentum. The forward-moving particle is the electron while the other directional particle is the proton. The exit caused the creation of finite masses in expanding space resulting in the first universe creation.

7.2 Proof

The proof of this theory could be based on calculated measurements and observed unexplained physical observations:

1. The speed of an electron in orbit in the hydrogen atom was found to be 9.9×10^6 m/s, which is a little faster than the speed of an electron as calculated in Bohr's hydrogen atom model.

2. The difference between the electron speeds, as calculated by Bohr and by this theory, is responsible for the heat energy possessed by the newly formed hydrogen atom.

3. The temperature of the newly formed hydrogen gas was calculated and found to be 1.58×10^6 degrees, which is in agreement with the temperature found in stars during their early formation.

4. Black holes were found to be gates to the speed-of-light dimension. It was found that energy cannot be observed inside them from rest and that they possess extremely high gravitational field. Black holes could be the source of particles sonic emission.

5. Electrostatic forces where related to the newly formed electrons and protons thrive to return their ground state, which is the speed of light.

6. Electrons and protons in orbit stay at a distance from each other in orbit in spite of the strong electrostatic attractive forces, limited by their increased masses as they approach the speed of light again.

7. This theory explained the evolution and nature of elementary and subelementary particles.

8. This theory is in agreement with universe expansion.

9. This theory supports the theory of a constant density of the universe.

This new theory is a simple and beautiful way to explain evolution and creation of the universe as predicted by Albert Einstein[24].

Chapter 8

Results and Conclusion

When we question the origin and evolution of the universe, the following points come to mind:

1. Where did this universe come from?
2. How were space and the elements created?
3. Why hydrogen is the most dominant element in the Sun?
4. Why are all stars overheated during their early formation?

This book answered all these questions and many more.

Observations taken in reference frames in other inertial reference frames differ from those taken inside a specific one. According to classical physics, constant-speed (inertial) reference frames can exist at any speed ranging from zero to infinity. The theory of special relativity agrees with the existence of inertial reference frames at any speed but sets a constraint: inertial reference frames cannot accelerate to the speed of light because of the increased mass restriction near that speed. This constraint does not include the existence of an inertial at the speed of light.

For an observer at rest, mass, length, and time measurements in a traveling reference frame near the speed of light change into heavier mass, shorter length, and dilated time, contrary to classical physics, where all measurements are invariant. These measurements remain invariant as they were at rest for the observer inside the traveling reference frame. This theory is explained mathematically by Lorentz transformation equations. According to the theory of special relativity, measurements of mass, length, and time depend on the speed of the measured values in reference to the speed of the measuring party. Inertial reference frames differentiate the speeds of others. When the traveling reference is at the speed of light, space contracts to zero, time stops, and mass is infinite or undefined as observed from rest; these measurements are according to Lorentz transformations. In other words, a reference frame traveling at the speed of light cannot be detected by an observer at rest. Accordingly, in this work, the origin of this universe started at the speed of light.

Two parallel universes equal in mass and size in two different dimensions exist; one universe is our universe while the other exists at the speed of light. Because the two dimensional universes are traveling at the speed of light in reference to each other, according to Einstein's theory of special relativity and Lorentz transformations, neither universe can observe the other because of space contraction and undefined mass at the sonic speed. The two universes are unable to detect each other, but they exist side by side.

Contrary to the big bang theory, instead of being created in extremely short time, our universe was created in bits and pieces continually basis in the other dimension or at the speed of light. One tiny particle exiting the speed-of-light reference frame could have created our empty but expanding space. Following expanding universe tiny particles started flowing from the other dimension filling the created empty space.

The tiny particles exiting the sonic reference frame are no more than electrons and protons exiting from the other dimension in equal and opposite momentum. Protons are ejected in the opposite direction to the motion of the other dimension, which is moving at the speed of light. Therefore, protons obtain speeds slower than light. Electrons are ejected in the same direction as the speed of light, thus obtaining speeds faster than light. Both electrons and protons attempt to recombine to return to their ground state, which is the speed of light. These particles are unable to reach the speed of light again, because their masses become infinitely heavy requiring infinite energy to break this speed barrier. Instead, the electron and the proton revolve around each other, trying to compensate for their excess energy forming the hydrogen atom, the basis of all matter. After the electron and proton attachment, they still possess excess energy, which is converted to heat energy that overheats the newly formed hydrogen atom. Mathematical analysis according to this proposed model showed that electron speed after exiting the speed of light, as measured at rest is only little higher than the electron speed as predicted by Bohr's model. The energy difference is converted into heat energy. The temperature of the newly formed hydrogen gas was found to be in the range of the Sun's core temperature. The newly formed overheated hydrogen gas attracts each other by the action of gravitational attraction resulting in the early formation of stars in a newly formed space.

Because of this proposed model, several observed but unexplained events are explained in the context of this theoretical approach. The most interesting result is the explanation of intense heat energy found in newly formed stars. The existence of elementary particles with partial mass and charge was also explained, including particles with momentum but without mass.

It was also found that resting observer detects infinite mass and cannot observe any movement or energy in a sonic reference frame. As a result, a sonic reference frame is characterized by dark appearance and infinite gravitational field. This is the same quality found in black holes. Black holes are thought to be gates or entrances to the other dimension.

8.1 Summary

Prior to 13.7 billion years ago, the age of the universe as predicted by the big bang theory, the universe was only a point in space traveling at the speed of light. In the beginning, two equal momentum particles were ejected from that point in the direction of motion of the speed of light. One particle obtained a speed lower than light (subsonic) while the other obtained speed faster than light (supersonic). According to Lorentz transformations, a generalized complex space expanded and housed both particles that could not be observed previously is now observed from rest. The subsonic particle's mass decreased as a real, positive value while the supersonic particle's mass also decreased but as negative imaginary value.

The supersonic particle was found to be the electron, with the negative sign indicating the electrostatic charge as compared with the subsonic proton's positive charge. The imaginary term associated with the supersonic mass signifies the difference between the two ejected masses, the electron and the proton.

The two ejected particles in an effort to return to their ground state, which is the speed of light, attract each other, thus explaining the electrostatic attraction property. Because both particles change speeds as they approach the speed of light again, their masses increase according to the Lorentz mass transformation. Exhausting all their available energy, they stop at a certain distance limited by their mass increase, defining the atomic radius in the hydrogen atom. The electron supersonic speed was calculated and found to be 1.8175×10^{10} m/s as measured from the supersonic reference frame, in reference to the proton's zero speed.

Applying the Lorentz velocity transformation, the electron speed was found to be 9.9×10^{6} m/s, as measured from rest. This resting speed is in the order of, but a little faster than, the electron's speed as calculated by Bohr. The difference between the two speeds is suggested to be the heat energy possessed by the newly formed hydrogen atom. The heat energy was calculated and found to be $1.58 \times 10^{6}\,°C$. This temperature is found in the Sun's core. Also, hydrogen is the main element on the Sun before the fusion process. The newly created overheated hydrogen cluster together by the action of gravity forms stars and galaxies.

The existence of another dimensional universe equivalent to our universe, but unobservable from rest was shown to exist in parallel with our universe. It was also shown that a resting observer detects infinite mass and cannot observe any movement or energy in the other dimension. Black holes have this quality; therefore, they are believed to be gates to the other dimension. The particles' emission, as explained earlier, are believed to flow off from black holes at a random but continuous process in all directions similar to dust particles flowing off a speeding car on a dusty road. A great bulk of those particles line up to form electrons and protons; others form particles with partial mass and charge and some with momentum only.

8.2 Main Results

The main hypothesis in this work is the evolution of electrons and protons from the speed of light forming the early hydrogen atom, the origin of the universe. This hypothesis showed some truth when the calculated speed of the electron was in the order of Bohr's electron speed in orbit. The calculated speed of the electron in its final excursion was found to be little faster than the electron speed in Bohr's orbit. The difference between the two speeds was explained to be the kinetic energy that is converted into heat energy. The calculated temperature of the newly formed hydrogen gas was found to be equivalent to the temperature in the Sun's core. These results are strong indications of the validity of this theory.

The evolution of the electrons and protons is coupled with the inflation of the universe space.

The most interesting findings in this work are the existence of two dimensional or two parallel universes existing side by side and that they are traveling at the speed of light in reference to each other. This was found as a direct application of Lorentz transformations.

This theory predicted the evolution of elementary particles with partial mass and charge and particles with momentum only.

The author believes that this theorized work can open the door for extensive future research that can explain many unexplained observations.

8.3 Conclusion

The most important result in this work is the existence of another dimensional universe the size of our universe. This universe is traveling at the speed of light in reference to our universe, which is set to be at rest.

Black holes were found to be also traveling at the speed of light. It was shown that energy cannot be observed inside a black hole together with extremely strong gravitational field as observed from rest. It is believed that black holes are entrances to the other dimensional universe. Dust particles flow from black holes, forming elementary and subelementary particles.

Elementary particles are the electrons and protons that flow in and against the direction of the motion of the sonic system while the subelementary particles flow in a direction lateral to the direction of the sonic system.

Electrons and protons propagate in the formed space. In an effort to return to their ground state, which is the speed of light, they maneuver to recombine, but limited by their increased masses as they approach the speed of light. They stop at a certain distance that we may call it the atomic radius. The result is an overheated hydrogen atom.

8.4 Recommendations for Future Research

Many topics can be investigated in the context of this work; the following could be suggested as a future research:

1. In the preceding development of the evolution of elementary particles and the hydrogen atom, it was assumed the existence of two reference frames, one traveling at the speed of light in reference to the other stationary system. On the other hand, the stationary system could also be traveling at the speed of light in reference to the other system. The question is, "How have these two systems been moving for so many billions of years?"

2. From Bohr's model the distance between the proton and the electron in orbit is about (5.3×10^{-11} m). In the above model of the evolution of the hydrogen atom, this radius was assumed to be the distance that separates the electron and the proton, beyond which the relativistic masses of both particles have consumed all the energy they obtained during their initial exit from the sonic system. The questions are, "Why this distance and how can it be calculated?"

3. As suggested, universe expansion needs further investigation in the context of this work.

4. A good point to investigate is how and why the neutron formed.

5. Physics can be investigated in the framework of this proposed theory.

Appendix A

Maxwell Equations

In 1873, the Scottish physicist and mathematician James Clerk Maxwell developed his four famous equations from historical experimental laws of static and, later, dynamic (time-varying) electric and magnetic fields with the aid of mathematical developments in vector calculus.

Maxwell assembled previous electric and magnetic laws based on experimental observations and put them in the form of four partial differential equations. These laws are Gauss's laws in both electric and magnetic fields and Ampere's and Faraday's laws. The solutions of these equations laid the foundations for today's wireless communication.

One of the most interesting results showed that light is an electromagnetic wave traveling at a constant speed in free space based on two universal constants the relative permittivity and permeability of free space. Albert Einstein fascinated with this result; he used this result to introduce his theory of special relativity.

The following is a rundown of the development of these equations.

A.1 Gauss's Law in Electric Fields

Gauss states that "the closed surface integral of the elemental electric field vector over a closed surface vector element d**S** (dot product) equals the charge enclosed by that surface divided by the relative permittivity."

$$\oint \boldsymbol{E}.\,d\boldsymbol{S} = \frac{Q}{\varepsilon_0} \qquad (A.1)$$

The enclosed charge Q equals the volume integral of the volume charge density over the elemental volume dV:

$$Q = \int \rho_v \, dV \qquad (A.2)$$

Combining equations (A.1) and (A.2),

$$\oint \boldsymbol{E}.\,d\boldsymbol{S} = \frac{Q}{\varepsilon_0} = \frac{1}{\varepsilon_0}\int \rho_v \, dV \qquad (A.3)$$

Where:

E is the electric field vector,

dS is the elemental surface area,

Q is the electric charge enclosed by the surface S,

ρ_v is volume charge density enclosed by the surface S, and

ε_0 is the permittivity of free space and is equal to $10^{-9}/(36\pi)$,

The permittivity of free space is a universal constant.

In equation (A.3), the integral on the left is over the closed surface S, the volume integral on the right is the volume enclosed by the same surface S.

The divergence theorem states that any closed surface integral of a vector is equal to a volume integral bounded by that surface. Thus, for any vector A, the divergence theorem states that

$$\oint A.\,dS = \int (\nabla.A)\,dV \qquad (A.4)$$

Again, in equation (A.4), the volume integral on the right-hand side is the same volume enclosed by the surface S. The proof of the divergence theorem in (A.4) can be found in any advanced calculus book. Inserting the vector E instead of the vector A in equation (A.4),

$$\oint E. dS = \int (\boldsymbol{\nabla}. E) \, dV \qquad (A.5)$$

Combining equations (A.3) and (A.5),

$$\oint E. dS = \frac{1}{\varepsilon_0} \int \rho_v \, dV = \int (\boldsymbol{\nabla}. E) \, dV \qquad (A.6)$$

Because the volume integrals in equation (A.6) represent the same volume; equation (A.6) could be reduced to

$$(\boldsymbol{\nabla}. E) = \frac{\rho_v}{\varepsilon_0} \qquad (A.7)$$

Equation (A.7) is Maxwell's first equation as derived from Gauss's law in electric fields, where again ρ_v is the charge density and ε_0 is the permittivity of free space.

A.2 Gauss's Law in Magnetic Fields

In similar analysis as presented in section A.1, for any magnetic field vector **H**, Gauss's law becomes

$$\oint \boldsymbol{H}.d\boldsymbol{S} = \frac{Q_m}{\mu_0} = \frac{1}{\mu_0} \qquad (A.8)$$

Where:

 H is the magnetic field intensity,
 Q_m is magnetic charge,
 μ_0 is the permeability of free space, and
 ρ_{vm} is the magnetic charge density.

The left surface integral encloses a surface **S** while the right volume integral is carried out over the volume enclosed by the same surface **S**. Using the divergence theorem again in a similar analysis as for the electric field shown earlier, equation (A.8) reduces to

$$(\boldsymbol{\nabla}.\boldsymbol{H}) = \frac{Q_m}{\mu_0} \qquad (A.9)$$

Because the magnetic charges always exist in pairs, north and south charges, and because they are of opposite polarities, their sum inside the closed surface is always equal to zero. Equation (A.9) reduces to

$$(\nabla.H) = 0 \qquad (A.10)$$

Equation (A.10) is Maxwell's second equation as derived from Gauss's law.

A.3 Ampere's Law

Ampere's law states that the closed line integral of the magnetic field vector **H** around a closed path is the same as the net current enclosed by the closed path, or

$$\oint H.dl = I_{enc} \qquad (A.11)$$

Where:

    H is the magnetic field intensity,
    dl is the elemental current length along the line path, and
    I_{enc} is the total current inside the path.

Stokes's theorem states that for any vector **H**, the line integral of **H** multiplied by the elemental line length is equal to the curl of **H** multiplied by the area enclosed by that line:

$$\oint H.\, dl = \int (\nabla \times H).\, dS \qquad (A.\,12)$$

The term I_{enc} in equation (A.11) is equal to the surface integral of the current density J over the area S, or

$$I_{enc} = \int J.\, dS \qquad (A.13)$$

Because S is the same area in equations (A.12) and (A.13), then equations (A.11), (A.12), and (A.13) can be equated, resulting in

$$\oint H.\, dl = \int (\nabla \times H).\, dS = \int J.\, dS \qquad (A.\,14)$$

Then

$$(\nabla \times H) = J \qquad (A.\,15)$$

Where:

> H is the magnetic field intensity and
> J is the total current density.

To investigate the current density in equation (A.15), we can generalize that any material can be electrically represented as a parallel combination of a resister and a capacitor. The total current I flowing in the material is the sum of the current passing through the resistor, called the conduction current, I_c, and the current passing through the capacitor, called the displacement current, I_d. The total current in the material is the sum of its conduction and its displacement current; thus,

$$I = I_c + I_d \qquad (A.16)$$

From Ohm's law, the conduction current is

$$I_c = \frac{V}{R} \qquad (A.17)$$

where *V* *is* the voltage (potential) between the two ends of the material with resistance *R;* the voltage in (A.17) can be written as

$$V = \boldsymbol{E}d \qquad (A.18)$$

Where:

E is the electric field between the two ends of the material,

d is the resistance length

From circuit theory

$$R = \frac{d}{\sigma S} \qquad (A.19)$$

Where:

d is the length of the material, same *d* as in (A.17) above,

σ is the conductivity of the material, and

S is the cross-sectional area of the material normal to the electric field.

Substituting (A.18) and (A.19) into equation (A.17) and setting σ$\boldsymbol{E}$ = $\boldsymbol{J}_c$, which is Ohm's law at a point:

Ghassan Hanna Halasa

$$I_c = Ed\frac{\sigma S}{d} = \sigma ES \qquad (A.20)$$

Redefining the conduction current in terms of the current density vector in the integral form; equation (A20)a can be written as

$$I_c = \sigma \int \boldsymbol{E}.d\boldsymbol{S} = \int \boldsymbol{J}_c.d\boldsymbol{S} \qquad (A.21)$$

Where; the integral to the right of equation (A.21) is the conduction current in terms of the conduction current density vector $\boldsymbol{J}_c$.

Because the two integrals on the right-hand side of (A.21) are over the same area $\boldsymbol{S}$, then

$$\boldsymbol{J}_c = \sigma E \qquad (A.22)$$

Equation (A.22) is Ohm's law at a point.

As for the displacement (capacitive) current, circuit theory tells us that the current in a capacitor is the rate of change of the voltage across the capacitor with respect to time multiplied by the capacitance:

$$I_d = C\frac{\partial V}{\partial t} \qquad (A.23)$$

where C is the capacitance of the material and V is the voltage across the capacitor.

The capacitance of any parallel plate capacitor (where the ends of the material in question define a parallel plate capacitor) is defined as

$$C = \frac{\varepsilon S}{d} \qquad (A.24)$$

where S is the cross-sectional area of the material normal to the current flow and d is the length of the material.

Combining equations (A.24) and (A.23) and (A.18),

$$I_d = \frac{\varepsilon S}{d} \cdot \frac{\partial (Ed)}{\partial t} = \varepsilon S.\frac{\partial E}{\partial t} \qquad (A.25)$$

In the integral form, equation (A25) can be written as

$$I_d = \varepsilon \frac{\partial E}{\partial t} \int d\,S = \int J_d\,dS \qquad (A.26)$$

Therefore,

$$J_d = \varepsilon \frac{\partial E}{\partial t} \qquad (A.27)$$

Equation (A.16) becomes

$$I = I_c + I_d = \int J.\,dS = \int J_c.\,dS + \int J_d.\,dS \qquad (A.28)$$

Because all the surface integrals in (A.28) are over the same area, then

$$J = J_c + J_d \qquad (A.29)$$

Then equation (A.29) becomes

$$J = \sigma E + \varepsilon \frac{\partial E}{\partial t} \qquad (A.30)$$

Inserting equation (A.30) into (A.15) yields

$$(\nabla \times H) = J_c + J_d = \sigma E + \varepsilon \frac{\partial E}{\partial t} \qquad (A.31)$$

Or

$$(\nabla \times H) = \sigma E + \varepsilon \frac{\partial E}{\partial t} \qquad (A.32)$$

Equation (A.32) is Maxwell's third equation as derived from Ampere's law.

A.4 Faraday's law

Faraday found that when a conductor is placed in a time-varying magnetic field a voltage is induced across the ends of the conductor. Faraday's law states that the induced voltage in one conductor is equal to the negative rate of change in the magnetic flux:

$$V_{ind} = -\frac{d\Psi}{dt} \qquad (A.33)$$

Where:

Ψ is the magnetic flux (function of time) and
V_{ind} the induced voltage in the conductor.

The flux, Ψ, may be written in terms of the magnetic flux density vector, **B**, multiplied by the surface area, where the area vector is normal to the surface and hence parallel to the flux flow:

$$\Psi = \int \boldsymbol{B}.\,d\boldsymbol{S} \qquad (A.34)$$

The magnetic field density **B** is equal to the magnetic field intensity **H** multiplied by the relative permeability μ, (**B** = μ**H**), equation (A.34) becomes

$$\Psi = \int \boldsymbol{B}.\,d\boldsymbol{S} = \mu \int \boldsymbol{H}.\,d\boldsymbol{S} \qquad (A.35)$$

where μ is the relative permeability.

Inserting (A.35) into equation (A.33),

$$V_{ind} = -\mu \frac{\partial}{dt} \int \boldsymbol{H}.\,d\boldsymbol{S} \qquad (A.36)$$

From Coulomb's law, the electric potential between two points, A and B, in static electric fields is defined as

$$V_{AB} = -\int_{A}^{B} \boldsymbol{E}.\,d\boldsymbol{l} \qquad (A37)$$

where dl is an elemental length of the conductor.

For a closed path of integration, the potential is then

$$\oint \boldsymbol{E}.d\boldsymbol{l} = 0 \qquad\qquad (A.38)$$

For a closed loop placed in a time changing magnetic field, and according to Ampere's law, equation (A.37) is no longer equal zero; instead, it is equal to the right-hand side of equation (A.36):

$$\oint \boldsymbol{E}.d\boldsymbol{l} = -\mu\frac{\partial}{dt}\int \boldsymbol{H}.d\boldsymbol{S} \qquad (A.39)$$

From Stokes's theorem, the line integral to the left of (A.39) can be changed into surface integral:

$$\int (\boldsymbol{\nabla} \times \boldsymbol{E}).d\boldsymbol{S} = -\mu\frac{\partial}{dt}\int \boldsymbol{H}.d\boldsymbol{S} \qquad (A.40)$$

Because the surface integral is the same on both sides of equation (A.40), the equation becomes

$$(\boldsymbol{\nabla} \times \boldsymbol{E}) = -\mu \frac{\partial H}{dt} \qquad (A.41)$$

Equation (A.41) is the fourth and last of Maxwell's equation.

A.5 <u>Solutions of Maxwell's Equations</u>

Maxwell's equations may be summarized as

$$(\boldsymbol{\nabla}.\boldsymbol{E}) = \frac{\rho_v}{\varepsilon_0} \qquad\qquad \text{(A.7)}$$

$$(\boldsymbol{\nabla}.\boldsymbol{H}) = 0 \qquad\qquad \text{(A.10)}$$

$$(\boldsymbol{\nabla} \times \boldsymbol{H}) = \sigma E + \varepsilon \frac{\partial E}{\partial t} \qquad \text{(A.32)}$$

$$(\boldsymbol{\nabla} \times \boldsymbol{E}) == -\mu \frac{\partial H}{dt} \qquad \text{(A.41)}$$

In free space, the permittivity, permeability, conductivity, and volume charge density are:

$$\varepsilon_0 = \frac{10^{-9}}{36\pi}$$

$$\mu_0 = 4\pi \times 10^{-7}$$

$$\sigma = 0$$

$$\rho_v = 0$$

Maxwell's equations then reduce to

$$(\boldsymbol{\nabla}.\boldsymbol{E}) = 0 \qquad (A.42)$$

$$(\boldsymbol{\nabla}.\boldsymbol{H}) = 0 \qquad (A.43)$$

$$(\boldsymbol{\nabla} \times \boldsymbol{H}) = \varepsilon_0 \frac{\partial \boldsymbol{E}}{\partial t} \qquad (A.44)$$

$$(\boldsymbol{\nabla} \times \boldsymbol{E}) = -\mu_0 \frac{\partial \boldsymbol{H}}{\partial t} \qquad (A.45)$$

Taking the del cross product to the both sides of equation (A.45),

$$\mathbf{V} \times (\mathbf{V} \times \mathbf{E}) = -\mu_0 \frac{\partial(\mathbf{V} \times \mathbf{H})}{\partial t} \qquad (A.46)$$

Using the vector identity,

$$\mathbf{V} \times (\mathbf{V} \times \mathbf{E}) = \mathbf{V}(\mathbf{V}.\mathbf{E}) - \mathbf{V}^2 \mathbf{E} = -\mu_0 \frac{\partial(\mathbf{V} \times \mathbf{H})}{\partial t} \qquad (A.47)$$

Substituting (A.42) and (A.44) into (A.47) and rearranging terms yields,

$$\mathbf{V}^2 \mathbf{E} - \varepsilon_0 \mu_0 \frac{\partial^2 \mathbf{E}}{\partial t^2} = 0 \qquad (A.48)$$

The homogeneous partial differential equation (A.48) represents a wave equation propagating, for example, in the z direction at speed u, with the electric field, $\mathbf{E}$, time varying in the x or y directions normal to the direction of propagation. The velocity, u, is equal to

$$u = \frac{1}{\sqrt{(\varepsilon_0 \mu_0)}} = 3 \times 10^8 \text{ m/s} \quad \text{(A.49)}$$

The speed in (A.49) is equal to the speed of light.

In a similar analysis, the magnetic field shows a similar result with a phase velocity as in (A.49). The magnetic and electric fields are found to be normal to each other. Light is then found to be an electromagnetic wave traveling at a constant velocity in (A.49).

Albert Einstein was fascinated with the results obtained in equation (A.49). He said that because both the relative permeability and permittivity of free space are universal constants, the speed of light must be universal constant, where everything else must vary according to that. The second postulate in Einstein's theory of special relativity is a direct outcome of equation (A.49).

Appendix B

Lorentz Transformations

B.1 Velocity Relation

Einstein's theory of special relativity is based on the assumption that all motion is related to a reference frame and that there are no absolute or preferred inertial reference frames. It is based on two postulates:

1. The laws of physics are the same for all inertial reference frames, where inertial reference frame is defined as a nonaccelerating reference frame.

2. Light always propagates through free space at a definite velocity, c, which is independent of the state of motion of the emitting body.

The first postulate is in full agreement with Newtonian and Galilean physics. This postulate states that a body subject to no external force will move with a constant velocity if it is in a nonaccelerating (inertial) reference frame. The second postulate is as a consequence and is in full agreement with the result obtained from the solution of Maxwell's equations. Equation (A.48) showed that light is an electromagnetic wave traveling in free space at a constant velocity, c, the speed of light. Because light is a wave, just like any other wave, its velocity is independent of the state of motion of the emitting body.

To derive Lorentz transformations let us assume two inertial reference frames S and S'. Coordinate system S is at rest with respect to coordinate system S' traveling to the right with a constant velocity, v. The x-axis of both coordinate systems is in the same direction, as shown in figure B.1.

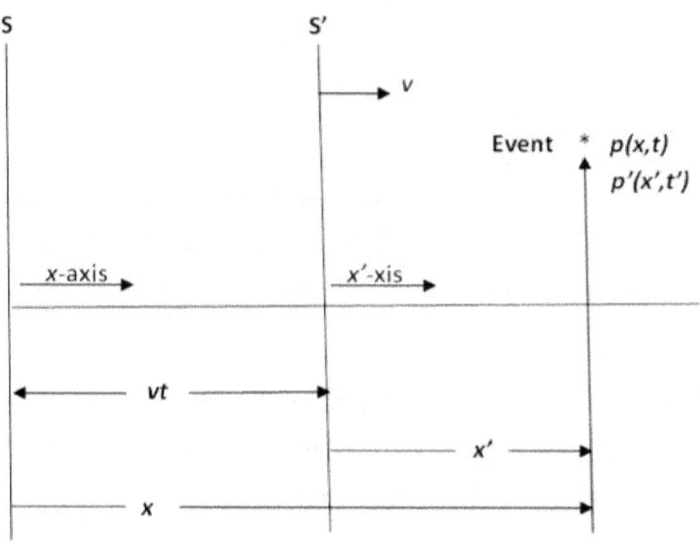

Figure B.1: The plot of two inertial reference frames, S is at rest, S' traveling with a velocity v, in m/s

At $t = 0$, both coordinates, S and S', coincide, and the clocks of both systems are synchronized. After a certain time, an event occurs at point p. The coordinates of the event p, according to S, are (x, t) whereas, according to S', they are (x', t'). According to classical mechanics, distance x as measured by the stationary system S is

$$x = x' + vt' \qquad (B.1)$$

and the distance x' as measured by the traveling system S' is

$$x' = x - vt \qquad (B.2)$$

Assume that event p is a flash of light. Because the speed of light, c, is invariant in any reference frame according to postulate two, then

$$x = ct \qquad (B.3)$$

$$x' = ct' \qquad (B.4)$$

To equate equations (B.1) and (B.2) with equations (B.3) and (B.4) respectively, then the right-hand sides of both equations (B.1) and (B.2) must be multiplied by some term, which is a function of v; thus,

$$x = ct = (x' + vt') f(v) \quad \text{(B.5)}$$

$$x' = ct' = (x - vt) f(v) \quad \text{(B.6)}$$

The term $f(v)$ is the same in both equations (B.5) and (B.6) because both equations are the inverse of each other. The function $f(v)$ can be found by eliminating $x, x', t, \text{ and } t'$, yielding

$$f(v) = \frac{1}{\sqrt{(1 - \frac{v^2}{c^2})}} = \gamma \qquad (B.7)$$

γ in equation (B.7) is called Lorentz factor.

B.2 <u>Length Relation</u>

Equations (B.5) and (B.6) become

$$x = \frac{(x' + vt')}{\sqrt{1 - \frac{v^2}{c^2}}} \qquad (B.8)$$

$$x' = \frac{(x - vt)}{\sqrt{1 - \frac{v^2}{c^2}}} \qquad (B.9)$$

Equations (B.8) and (B.9) are the Lorentz distance transformations.

For a ruler, the coordinates at the two ends according to S are (x_1, t_1) and (x_2, t_2) and (x_1', t_1') and (x_2', t_2') in S'. The actual length of the ruler as measured in S is Δx:

$$\Delta x = x_2 - x_1$$

$$\Delta t = t_2 - t_1$$

Substituting and subtracting in (B.8),

$$\Delta x = \frac{(\Delta x' + v\Delta t')}{\sqrt{1 - \frac{v^2}{c^2}}} \qquad (B.10)$$

and

$$\Delta x' = x_2' - x_1'$$

$$\Delta t' = t_2' - t_1'$$

Substituting into (B.9),

$$\Delta x' = \frac{(\Delta x - v\Delta t)}{\sqrt{1 - \frac{v^2}{c^2}}} \qquad (B.11)$$

For $t_1' = t_2'$ or $\Delta t' = 0$, equation (B.10) becomes

$$\Delta x = \frac{\Delta x'}{\sqrt{1 - \frac{v^2}{c^2}}} \qquad (B.12)$$

Rearranging terms in (B.12),

$$\Delta x' = \Delta x \sqrt{1 - \frac{v^2}{c^2}} \qquad (B.13)$$

Equation (B.13) represents length contraction.

B.3 Time Relation

Divide equation (B.9) by c, substituting for t' as in equations (B.4):

$$\frac{x'}{c} = t' = \frac{\left(\frac{x}{c} - \frac{v}{c}t\right)}{\sqrt{1 - \frac{v^2}{c^2}}} \qquad (B.14)$$

Replacing t in equation (B.14) with equation (B.3),

$$t' = \frac{(t - \frac{v}{c^2}x)}{\sqrt{1 - \frac{v^2}{c^2}}} \qquad (B.15)$$

Equation (B.15) is the Lorentz time transformation.

For two events $E_1'(x_1', t_1')$ and $E_2'(x_2', t_2')$ on S' corresponding to $E_1(x_1, t_1)$ and $E_2(x_2, t_2)$ on S. Substituting the events into equation (B.15) and subtracting E_2' from E_1', where

$$\Delta x = x_2 - x_1$$
$$\Delta t = t_2 - t_1$$
$$\Delta t' = t'_2 - t'_1$$

$$\Delta t' = \frac{(\Delta t - \frac{v}{c^2}\Delta x)}{\sqrt{1 - \frac{v^2}{c^2}}} \qquad (B.16)$$

For the same location, Δx equals zero, $\Delta t'$ is the time T' elapsed in S', and Δt is the corresponding time T elapsed in S; then equation (B.16) becomes

$$\Delta t' = \frac{\Delta t}{\sqrt{1-\frac{v^2}{c^2}}} \qquad (\text{B.17})$$

Or

$$T' = \frac{T}{\sqrt{1-\frac{v^2}{c^2}}} \qquad (B.18)$$

Equation (B.18) represents time dilation.

B.4 Velocity Relation

Dividing equations (B.11) by (B.16) and dividing the numerator and denominator of the right-hand side by Δt yields

$$\frac{\Delta x'}{\Delta t'} = \frac{(\frac{\Delta x}{\Delta t} - v)}{(1 - \frac{v}{c^2}\frac{\Delta x}{\Delta t})} \qquad (B.19)$$

Take the limits

$$lim_{\Delta t' \to 0} \frac{\Delta x'}{\Delta t'} = V' \quad and \quad lim_{\Delta t \to 0} \frac{\Delta x}{\Delta t} = V$$

Equation (B.19) becomes

$$V' = \frac{V - v}{1 - \frac{v}{c^2}V} \qquad (B.20)$$

Solving equation (B.20) for V in terms of V',

$$V = \frac{V' + v}{1 + \frac{v}{c^2}V'} \qquad (B.21)$$

Equations (B.20) and (B.21) are the Lorentz velocity transformations.

Appendix C

Relativistic Mass and Energy

C.1 <u>Mass Relation</u>

Momentum must be conserved in all reference frames

$$P = m\mathbf{v} = m'\mathbf{v}' \qquad (C.1)$$

Where the primed terms in equation (C.1) are in the S' reference frame as measured by an observer in S at rest. When Einstein was asked, "How about momentum?" he answered, "Multiply it by Lorentz factor." Deriving the relativistic mass is not an easy task. Some authors have divided the two velocities in (C.1); others have used different methods some used the quantum force relationship. One easy way to derive the momentum relation is if we assume that a ball at rest traveling along the y-axis. Because the moving reference frame is traveling along the x-axis, the observer measures the distance as y in the same way as the resting observer, but because of time dilation, the velocity seems to be larger; thus,

$$u_y = \frac{\Delta y}{\Delta t} \qquad (C.2)$$

$$u_y' = \frac{\Delta y}{\Delta t'} = \frac{\Delta y}{\Delta t \times \frac{1}{\gamma}} = \frac{\Delta y}{\Delta t} \times \sqrt{1 - \frac{v^2}{c^2}} \qquad (C.3)$$

Inserting (C.2) and (C.3) into (C.1) yields

$$m \times \frac{\Delta y}{\Delta t} = m' \times \frac{\Delta y}{\Delta t} \times \sqrt{1 - \frac{v^2}{c^2}} \qquad (C.4)$$

or

$$m = m' \sqrt{1 - \frac{v^2}{c^2}} \qquad (C.5)$$

and

$$m' = \frac{m}{\sqrt{1 - \frac{v^2}{c^2}}} \qquad (C.6)$$

m' in equation (C.6) is the relativistic mass.

C.2 Mass–Energy Relation

Elemental energy, in general, may be defined as the dot product of the applied force and the elemental distance

$$dE = \boldsymbol{F}.d\boldsymbol{x} \qquad (C.7)$$

where E is the energy and F is the force exerted for a distance dx.

According to Newton, force is defined as the rate of change of momentum, $\boldsymbol{P}$

$$F = dP/dt \qquad (C.8)$$

Inserting (C.7) into (C.8) yields

$$dE = dP\frac{dx}{dt} = vdP \qquad (C.9)$$

where v is the velocity.

The momentum P is defined as

$$P = mv \qquad (C.10)$$

Taking the derivative of both sides of (C10),

$$dP = mdv + vdm \qquad (C.11)$$

Inserting (C.11) into the right-hand side of (C.9),

$$dE = mvdv + v^2dm \qquad (C.12)$$

Equation (C.6) relates the speeding mass m' to the rest mass m; taking the derivative of m' with respect to v yields

$$\frac{dm'}{dv} = m\{(-\frac{1}{2})(1 - \frac{v^2}{c^2})^{-\frac{3}{2}}(-2\frac{v}{c^2}) \qquad (C.13)$$

$$\frac{dm'}{dv} = \frac{mv}{c^2}\left(1 - \frac{v^2}{c^2}\right)^{-\frac{3}{2}} \qquad (C.14)$$

Replacing *m* by *m'* by using in equation (C.6) yields

$$\frac{dm'}{dv} = \frac{m'v}{c^2} \times \frac{1}{1 - \frac{v^2}{c^2}} \qquad (C.15)$$

or

$$\frac{dm'}{dv} = \frac{m'v}{c^2 - v^2} \qquad (C.16)$$

Cross-multiplying (C.16) yields

$$dm'(c^2 - v^2) = m'vdv \qquad (C.17)$$

Rearranging terms

$$c^2 dm' - v^2 dm' = m'v dv \qquad (C.18)$$

Or

$$c^2 dm' = v^2 dm' + m'v dv \qquad (C.19)$$

The right-hand side of equation (C.19) is recognized as dE in equation (C.12). Therefore,

$$dE = c^2 dm' \qquad (C.20)$$

Integrating both sides of equation (C.20),

$$E = m'c^2 \qquad (C.21)$$

Equation (C.21) is the famous Einstein's mass–energy equation.

REFERENCES

1. Philip J. E. Peebles, *Principles of Physical Cosmology,* Princeton University Press, Princeton, N.J., 1993

2. A. Peacock, *Cosmology Physics* (Cambridge: Cambridge University Press, 2002);

3. "Primeval atom," *Wikipedia,* accessed July 11, 2010,.http://en.wikipedia.org/wiki/Primeval_atom;

4. Björn Feuerbacher and Ryan Scranton, "Evidence for the Big Bang," *Talking Origins Archive*, January 25, 2006, accessed July 10,2010,http://www.talkorigins.org/faqs/astronomy/bigbang.html#evidence.

5. Edwin Hubble, "A Relation between Distance and Radial Velocity among Extra-Galactic Nebulae," *Proceedings of the National Academy of Sciences* 15, no. 3 (March 15, 1929): (168-173)

6. Abbe G. Lemaitre, "A Homogenous Universe of Constant Mass and Increasing Radius Accounting for the Radial Velocity of Extra-Galactic Nebulae," trans. by permission from *Annales de la Societe Scientifique de Bruxelles*, Tome XLVII, Serie A, Premiere Partie, Royal Astronomical Society, provided by NASA Astrophysics Data System.

7. G. Lemaitre, "The Evolution of the Universe: Discussion," *Nature* 128 (1931): 699–701.

8. H. Bondi and T. Gold, "The Steady State Theory of the Expanding Universe," *Monthly Notices of the Royal Astronomical Society*, 108 (1948): 252.

9. Edward L. Wright, "Errors in the Steady State and Quasi-SS Models," last modified, January 30, 2010, http://www.astro.ucla.edu/~wright/stdystat.htm.

10. F. Hoyle, G. Burbidge, and J. V. Narlikar, "A Quasi-Steady State Cosmological Model with Creation of Matter," *Astrophysical Journal*, 410, no. 2, pt. 1 (1993): 437–57.

11. F. Hoyle, "A New Model for the Expanding Universe," *Monthly Notices of the Royal Astronomical Society*, 108 (1948): 372.

12. A. H. Guth, "Inflationary Universe: A Possible Solution to the Horizon and Flatness Problems," *Physical Review D*, 23, no. 2 (1981): 347–56;

13. A. Guth, "Was Cosmic Inflation the Bang of the Big Bang," *The Beamline* 27, no. 3 (1997): 14–21, accessed July 14, 2010, http://nedwww.ipac.caltech.edu/level5/Guth/Guth_contents.html.

14. Eric Lerner, "Big Bang Never Happened Home Page and Summary," accessed July 14, 2010, http://bigbangneverhappened.org/.

15. Ned Wright, "Welcome to Ned Wright's Cosmology Tutorial," accessed July 14, 2010, http://www.astro.ucla.edu/~wright/cosmolog.htm.

16. Björn Feuerbacher, Ryan Scranton, 'Evidence for the Big Bang', http://www.talkorigins.org/faqs/astronomy/bigbang.html, Posted January 25, 2006

17. Bilaniuk, Olexa-Myron P.; Sudarshan, E. C. George (May 1969). "Particles Beyond the Light Barrier", Physics Today 22 (5): pp43–51

18. http://evolutionfacts.com/Ev-V1/1evlch01a.htm#BIG BANG THEORY, Evolution Encyclopedia Vol. 1, 'Origin and Evolution of the universe', Evolution Facts Inc,PO BOX 300 – ALTAMONT, TN. 37301 USA,

19. Relativity: The Special and General Theory by Albert Einstein, translated by   Robert William Lawson
Part I – The Special Theory of Relativity
http://en.wikisource.org/wiki/Relativity: The Special and General Theory/Part I

20. "Black Hole," Wikipedia, accessed June 2013, https://en.wikipedia.org/wiki/Black_hole.

21. "Frame of Reference," Wikipedia, accessed June 2013, https://en.wikipedia.org/wiki/Frame_of_reference.

23. R. M. Fedler and R. W. Rousseau, Elementary Principles of Chemical Processes (New York: John Wiley & Sons, 1978).

24. Wikipedia, accessed Feb 16, 2014)
http://www.goodreads.com/quotes/1360334-i-have-deep-faith-that-the-principle-of-the-universe